INFORMATION NEEDS FOR WATER MANAGEMENT

Information Needs for Water Management

Jos G. Timmerman
Senior Researcher
Alterra Wageningen University and Research Centre
Earth System Science Group
Wageningen
The Netherlands

CRC Press is an imprint of the
Taylor & Francis Group, an **informa** business
A SCIENCE PUBLISHERS BOOK

CRC Press
Taylor & Francis Group
6000 Broken Sound Parkway NW, Suite 300
Boca Raton, FL 33487-2742

Cover Illustrations reproduced by kind courtesy of the author of the book, Jos G. Timmerman.

Printed on acid-free paper

International Standard Book Number: 978-1-4665-9474-6 (Hardback)

Visit the Taylor & Francis Web site at
http://www.taylorandfrancis.com

CRC Press Web site at
http://www.crcpress.com

Science Publishers Web site at
http://www.scipub.net

Preface

This book is based on my experiences with specifying information needs in a range of projects, the details of which were the basis of my PhD thesis finalized in 2011. Working on my thesis I realized the difference between working on a manuscript that needed to be scientifically sound and a book that can support people in their work. The idea grew after finalizing the thesis—that it would be useful to turn the work, a scientific endeavor, into a practical guide, applying the methodology developed in the thesis. Such a book would offer hands-on support to practitioners in developing a process to specify information needs. This book is the result of that idea.

This book basically has two purposes. One is to develop the reader's understanding of the role and use of information in decision making, in the context of water management. The other is to provide support, by means of a structured approach towards specifying information needs.

The first chapter in the book gives a general overview of the policy making process, the role of information in that process, and the mismatch between the information that is produced and the information that is considered necessary. The second chapter describes the information cycle; the process of producing information. It describes the phases in the cycle and the way it can be applied in developing and improving that process. The other chapters follow the logical steps in developing and implementing the first phase in the information cycle; the information needs a specification process. They describe the design of the process, an analysis of the water management situation as the first step in deciding policy objectives, the structured breakdown of policy objectives into information needs, and the final steps to be able to develop the following phases of the information cycle towards an overall information network. Each chapter starts off with a general description of the issue at hand, followed by a hands-on description of the steps to be undertaken. Moreover, to support the reader, each chapter begins with an overview of the content and lessons that follow. Each chapter also ends with a number of practical exercises to help the reader better understand the contents and gain experience in specifying information needs.

For detailed information on the subjects covered in the book, the reader may consult the list of references. Moreover, where relevant, I have included boxes that provide background information on specific topics. The boxes are, however, not necessary for proper understanding of the book.

I hope this book will show readers that it is possible to specify information needs in a structured way and therewith provide sufficient support to elaborate on the information needs in their organizations.

Jos G. Timmerman
Alterra Wageningen University and Research Centre
Earth System Science Group
Wageningen
The Netherlands

Acknowledgements

Writing a book is not the work of one person alone. I would not have been able to write this book without the guidance I received on my PhD Thesis from my supervisors Wim Cofino and Katrien Termeer from Wageningen University and Research Centre (The Netherlands), and Euro Beinat from University of Salzburg (Austria) and Zebra Technologies Corporation (Chicago, USA). The invaluable comments from Andrew Quin, Royal Institute of Technology (Stockholm, Sweden), who reviewed an earlier version of this book, have improved the work.

Contents

1

Introduction—Setting the Scene

This chapter deals with the role of information and monitoring in water management and stresses the importance of specifying information needs to improve the production of information. Starting off with the global water management challenges it will introduce the policy process and the role of information therein. It will also explain the nature of the water information gap and the implications this has for assessing policy makers' information needs. Understanding the policy process and the role played in it by information is essential to be able to link information production to this process. Finally, the chapter will discuss the nature of policy problems and how they can be structured. After studying this chapter the reader will be able to:

- Explain the nature of the water information gap
- Explain the policy making process and the policy life-cycle
- Explain the role of information in the policy making process

1.1 INTRODUCTION TO WATER MONITORING

1.1.1 The Global Water Management Challenge

Water is an essential natural resource with limited availability. There is plenty of water on the Earth and in the ground, but it is not distributed evenly over the Earth's surface and in time. Many people have too little water to grow their crops, while on the other hand floods frequently threaten lives and harvests. Also, the quality of the available water is often poor and access to safe drinking water is lacking. Besides this, natural systems like the Aral Sea and other examples are under severe pressure from (often competing) human demands[1-4]. Water management, or when put in a broader context, water governance[5, 6], should be able to manage these issues.

It is generally acknowledged that for the abovementioned reasons, water should be managed in a sustainable way[1, 7]. This denotes that water policy must aim at developing in such a way that the present needs are met without compromising the ability of future generations to meet their needs[8]. Integrated Water Resources Management (IWRM) is sometimes questioned for not being comprehensive enough[9] or too vague[10], but is nevertheless generally considered as the proper concept for achieving sustainable water management[5, 6, 11]. The role for policy makers in this is to reach out for sustainability by directing progress towards integrating economic, ecological and socio-cultural dimensions for all human activity[12]. Sustainability, in this sense, requires balancing a variety of needs including future needs that are not yet clear. This is the larger context of water policy and water management.

1.1.2 The Need for Information

Effective water resources development and management is not possible without adequate information and benefits when the quality of information is improved[13-16]. Information on relevant characteristics supports and guides decision makers to determine the best ways to proceed and is the basic source to evaluate the effects of specific policies[17, 18]. Vast effort is hence put into the collection and dissemination of environmental information, especially by governments and government-related institutions[19].

The need for information in the field of water quality management has steadily increased over time. Water quality management was virtually non-existent up to approximately 1850, by which time local environmental conditions deteriorated severely with foul smelling, deoxygenated water as a result of industrialization[20]. The rapid developments in urbanization and industry after World War II in western Europe led to deterioration of the quality of surface waters. New industrial processes emerged, that increased the standard of living. These same processes produced increasing streams of wastewater containing organic loads, heavy metals, dyes, etc., Studies into the water quality situation were conducted whenever locally problems occurred. However, slowly water managers came to realize that these problems became long lasting and omnipresent. Irregular studies were not enough to deal with the pollution and a structured system of information gathering was needed. Regular water quality monitoring was not established until the 1950's in the USA, the former USSR and in a few European countries and extended to Canada and most of western Europe in the late 1960's and 1970's[21].

Box 1.1 Syndromes of river changes[22]

Direct and indirect human pressures on aquatic systems over centuries have led to syndromes of river change. These syndromes can be cured, which takes decades for some and centuries for others. Meybeck[22] distinguishes the following syndromes:

- Flow regulation, that is generally achieved through the construction of dams and reservoirs and sometimes through water diversions. The ecological impacts of such extreme regulation can be important.
- River fragmentation, where river courses are interrupted by multiple dams and reservoir cascades that greatly limit their longitudinal connectivity or change in lateral connectivity due to channelization, levees and embankment construction. Such fragmentation also results in major changes of the aquatic biota particularly fish (e.g. migratory species) and intersticial fauna.
- Neo-arheism is the dramatic reduction of river flow due to consumptive use of water and/or to diversions, particularly resulting from irrigation, but it may also be related to urban water demand in some dry regions. The impacts of neo-arheism are mostly observed in the coastal zone which is no longer supplied with fresh water, thus changing the salt balance, with essential nutrients, organic matter and sediments.
- Sediment unbalance is the gradual or rapid change of sediment transfer, suspended matter or bedload in river systems due to land-use changes and to reservoir building. As a result, most of the sediment is redeposited in foothills, river beds and floodplains and, more recently, in reservoirs instead of the river mouth.
- The salinization as a result from the release of dissolved salts from industrial and urban sources. The salinization process may result in severe limitations of water uses.
- The chemical contamination is related to most human activities such as mining and oil extraction, industries, urbanization and transport, agriculture.
- Acidification of continental waters is related to atmospheric fallout of sulfuric and nitric acids.
- Eutrophication, an excess of algal development in water bodies due to nutrient enrichment. The nutrient unbalance can lead to anoxia, deteriorating the aquatic life.
- Thermal unbalance that may be caused by thermal pollution or, e.g., by reservoirs' operation. Its impacts on the aquatic biota may be important.
- Biological introductions that result mostly from rapid increase of fluvial transport and of ocean transport through ballast waters. Biological introductions may have dramatic impacts on water resources and on aquatic ecology.

Meybeck and Helmer[23] give an overview of some of the major pollution problems arising over the years in industrialized countries (also see Box 1.1). They show that fecal and organic pollution were important problems even before 1900. Salinization, metal pollution and eutrophication emerged before the 1950s. Around the 1950s and 1960s, problems with radioactive wastes, nitrate and organic pollutants became apparent. This illustrates that every now and then new problems are manifested. In addition to what Meybeck and Helmer describe until 1989, in the late 1990's, among others, problems arising from oestrogenouos substances, tributyltins, and drugs became apparent[24-26]. With each emerging environmental issue, water management becomes more complex and, connected to this, each new problem leads to new information needs. As a consequence, there is a continuous push towards increasing the water quality monitoring efforts.

Moreover, realization grew that water management was not only a quality issue but had to be integrated over the various functions and uses of water. A balance was needed between ecological, economic and social issues[27, 28]. This is also where IWRM as a concept developed. In this way water monitoring developed over a few decades from measuring a few simple parameters into a complex process where many different parameters are measured in various frequencies on various locations.

Water quality monitoring nowadays is based on the supposition that water management is not possible without adequate information (Box 1.2) and that information should help decision makers arrive at sounder, faster and more transparent decisions[17]. The supposition is that information assists decision makers in rationalizing the choices made, that is, that the choices are based on objective criteria such as the minimization of costs or the maximization of benefits[35]. The overall premise of this book is that decision makers need information to help them make better informed decisions. Reality is usually more complex than that and this book will also try to describe possible deviations from the theory to a certain extent, based on available literature (Box 1.3).

Box 1.2 Access to environmental information

Access to environmental information is generally considered to be a fundamental right[29-32]. The US National Environment Policy Act (NEPA) already in 1969 explicitly stated that information regarding environmental problems must be made available to States, Counties, Municipalities, Institutions, and other entities, as appropriate[18, 33]. Free access to environmental information is however no absolute right; there are an extensive number of exemptions due to, for instance, security issues. And there are practical impediments like accessibility of, especially older, data. With digital archives access to data becomes easier, but the free access to information is now sometimes purposely turned into flooding people with information. De Villeneuve[34] gives a more detailed discussion on the benefits and limitations of access to environmental information.

Box 1.3 The importance of long-term studies

A review by the European Environment Agency (EEA)[36] of 14 major environmental issues in the late 20th century indicated that long-term environmental studies had a key role to play in the identification of environmental problems. The long-term data series were able to point out the problems and policies and measures could be developed accordingly. Although changes in policy often take longer to achieve than environmental scientists believe wise, the examples from the report show the importance of maintaining a wide range of long-term monitoring studies in order to identify emerging issues. Moreover, many contemporary environmental issues are global in scale and these require more coordinated and harmonized approaches to long-term monitoring.

1.1.3 Impact of Information

Talking about information also implies disseminating and communicating the information. Denisov and Christoffersen took an approach towards information in policymaking, of looking at the impact of information on decisionmaking[37]. They state that decisions are not only made on the basis of individual and institutional considerations, but are strongly influenced by visible and hidden systems of interests. The role of information is to help promote, develop and establish more formal management frameworks that are supposed to modify and steer the behavior of people or organizations in the desired direction, such as laws or economic mechanisms.

There are several steps (Box 1.4) to take before information can create an impact. The information produced is generally brought out in the form of maps, graphics, books, papers, etc. The way this information is communicated—through journals, the Internet, mass media or conferences—is important to reach a specific target group, like scientists, decision makers or the public at large. This target group can develop their own ideas based on this information. These ideas, in turn, can lead to changes in, for instance, laws and policies, or even values. Such changes can then lead to altered behavior which could finally result in better quality environment[37].

Box 1.4 The flow of information in water management policy

The tasks in water management and the flow of information between these tasks has been identified by Van Bracht[38]. In the process of developing water management policies, social, economic and ecological considerations are identified.

The demands of interest groups like drinking water suppliers, industry, agriculture, recreation, transport (navigation), and nature conservation are examined. Potential uses of the water system are determined by the properties of that water system. The policy, together with the demands of the interest groups and the potential uses determine the assignment of functions to the water system. These functions are limited by the societal constraints as put down in the policy, the demands the uses put on the water system as depicted by the interest groups and the limitations as set by the properties of the water system.

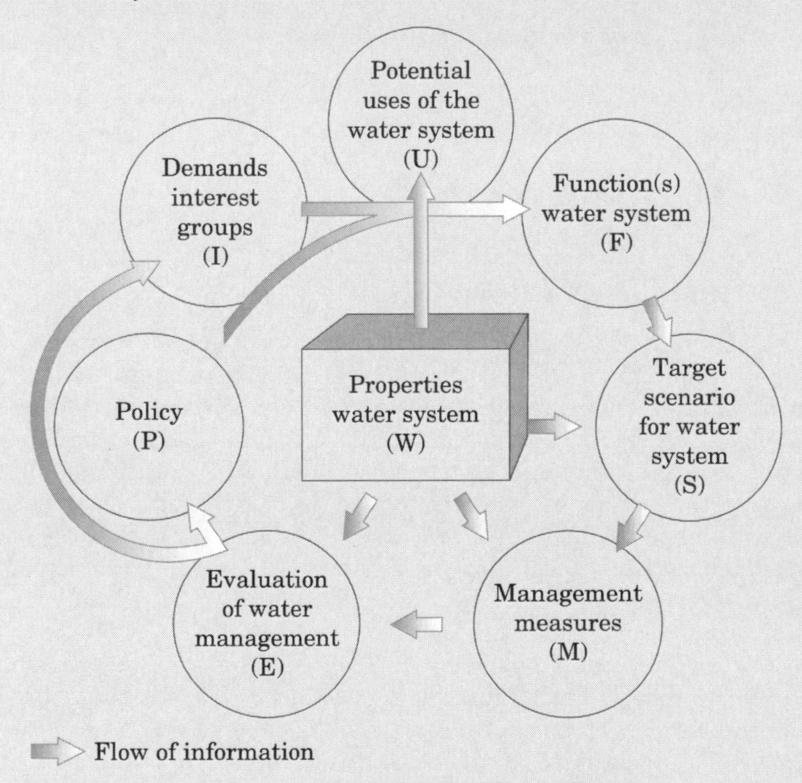

Figure 1.1 Flow of information in the water policy process

Derived from the functions, together with information on the properties of the water system, target scenarios can be developed that represent the conditions under which the water system can satisfy the demands as set by the assigned functions. These scenarios provide information for management measures like technical measures (for instance, construction of a weir) or legal measures (like imposing levies). Again these measures should be tuned to the properties of the water system. The water management will be evaluated by checking whether the measures have had the desired effects. This requires comparing the (changes in) properties of the water system to the management measures made.

The evaluation now provides new input to the policies and interest groups to start development of new water management and the cycle starts again[38, 39]. The entire flow of information as described here is coordinated from the information environment of the water management organization. Information to support environmental decisions in this framework should cover a wide range of issues and originates from various sources. Information from each of these sources is complex, often abundant, contradictory, having a differing quality, and is focused on scientific use rather than political use, as a result of which it is hard to access by policymakers or the general public. Information management is concerned with coordination and sharing of information, and translating this variety of information into a coherent set, and is as a result an important function in water management.

If information is to be effectively disseminated, it is apparent that good communication strategies need to be developed, linking to the audience that is to be reached. Creating effective communication tools therefore requires a strategy that can be developed by asking and answering a series of basic questions[40]:

- What is the objective? What should be achieved with this information? Is it raising or increasing of awareness? Is this information intended to change behavior or is it meant to start a dialogue with the public?[41].
- What is the target audience? Should the information focus on experts, on policymakers, or on lay people?
- What is the overall message that is to be conveyed? It is generally useful to frame the message in terms of problem and solutions.
- What format will be used to convey the message? Will it be a report, a presentation, a video, a conference, or maybe a combination of these?
- How will the message be distributed? Is it by means of a standard mailing list, by hooking up with the mass media, by dumping it all on a webpage, or by using social media?
- How will the success of this strategy be evaluated? Is it done by counting the number of references included in literature, quotes in the mass media, or should there be a notable change in opinion among the public at large?

Let us take a closer look at some of these questions. One essential question is: what is the message that is to be conveyed? Messages focusing only on the magnitude of an environmental problem (for instance, the state of the environment) or only on possibilities available for environmental action (for instance, new technologies) are generally less effective than messages that present both a problem and a solution in an interconnected manner. This enables the presented information to become more meaningful where information users perceive the pattern and also helps them to make changes (that can make information a constitutive force, see Section 1.2.9). Information on a problem, therefore, has to be produced in close co-operation with policymakers who can include solutions, On the other hand, in real life it may very well happen that information is used to justify decisions after, rather than before they are taken (information to direct decision-making, see Section 1.2.9)[37].

If information becomes a 'perception of pattern' or a 'constitutive force' (also see Section 1.2.9) it is capable of generating discussions. Moreover, information can act to generate processes whose value for decision-making at the end may be greater than that of the original 'catalyzing' information. This specifically happens when exchange of information is part of a participatory process where decision makers and other stakeholders are both closely involved in the information development process. Three effects of a participatory approach can be discerned[37]:

- Strengthened capacities of process participants to generate and handle environmental information. Networks of organizations and individuals involved in the process of collecting, processing, and producing information form a solid basis for production of information in the future. The people involved are aware of the purpose of the activity and are able to influence the collection of information to include their own interests and will therefore have more confidence in the information produced;
- Improved quality and acceptance of generated information due to multi-lateral inputs and controls. The participatory process means that information is subjected to quality control all through its production and not only at the end. As participants' varying perspectives are incorporated into the process , the resulting information reflects diverse values and interests, thus making the final product more salient (see Section 1.2.2);
- Better awareness of the findings among process stakeholders due to their direct involvement in the process, thereby attaching higher value on the findings and higher acceptance of the consequences of the findings. This also has influence on a wider audience as the stakeholders that participated have an inherent interest in broadcasting the results of the process to their acquaintances. Incorporation of major decision makers is likely to influence policy decisions.

Changing the information behavior - that is how individuals approach and handle information including searching for it, using it, modifying it, sharing it, hoarding it, even ignoring it - of an organization[42] towards a participatory approach and better cooperation, strengthens these effects. People in an organization that are used to such a participatory, social learning approach will be more open to the exchange of views and opinions and will be more aware of the process.

What determines the success of the communication strategy? Assessing the overall impact of information can be difficult for two reasons[37]:

1. Information itself, in the form of raising awareness among the public at large, is a relatively weak instrument of environmental

management compared to more efficient policy instruments for changing human behavior, such as laws, taxes, licences, or voluntary agreements. Information makes its strongest impact when it catalyzes, supports and relies upon more powerful mechanisms. On the other hand, enforceability of environmental policy instruments, especially licensing, is often dependent on information; well-informed people will be more likely to understand and implement the (licensing) requirements. The policy instruments are, however, often designed without much forethought about monitoring capabilities, thus crippling the evaluation possibilities and therewith their applicability[43].

2. Both short-term and long-term effects exist and need to be considered. Pronounced short-term effects are fairly easy to observe; long-term effects on the other hand are less easily detected and even less liable to be connected to the initial information. In general, even knowledge about tangible short-term effects still remains sporadic, also as often monitoring of the effects is generally not included.

Then, there are questions of what format to use and how to distribute the message. Based on knowledge of the needs and capabilities of the intended users, the information can be presented in such a way that it best suits their interests and expectations [37]. However perfect the quality and presentation of the information, it also needs to be communicated in a way that connects to the audience to attain maximum impact. Moreover, active communication is more effective than passive reporting to influence decision-makers; there is consequently a need to actively communicate the results. But too active communication may be perceived as pushy, that in turn may result in lower instead of higher impact.

The above is only a summary of issues that play a role in communicating information with the aim of making an impact. It will not be further discussed in this book, but ideas developed on the subject by many scholars can be found in literature. The important message from this section is that a participatory approach is needed, even when starting a process of information production, to ensure that the proper questions and issues are raised and addressed and that the resulting information is better and more widely accepted. And with that, resulting changes in behavior may be better accepted.

1.1.4 Developments in Dutch Water Management

The Netherlands has a long history of water management. The first dikes and dams were built in this country over 2,000 years ago. Organization of water management has changed over the past 500 years from numerous local administrations dealing with small areas to a complex of several layers of government on national, provincial, and regional levels (Box 1.5), co-operating and competing over water management[44, 45]. The developments in Dutch water quality management are briefly described in this section as an example of how water management has evolved over the last decades into a complex policy issue.

Box 1.5 Water Boards in The Netherlands

By the end of the 18th century over one thousand official and relatively independent bodies (Water Boards) were responsible for water management in The Netherlands, making water management a very local issue. This scattering of responsibilities and independency could lead to dangerous situations. For instance, maintenance of the sea defense dike was the responsibility of only a small community adjacent to the sea. Inland polder communities would not contribute to the maintenance of these larger dikes although their safety was also dependent on these sea dikes; the smaller inland dikes would not hold if the sea dikes would break. This situation could appear fatal during storms[46] and a major reason for Water Boards to merge with other boards to form larger boards that could more easily take care of the higher construction and maintenance costs of the sea defense.

Later on arguments like improved efficiency and effectiveness became more important reasons for mergers of Water Boards. This has led to a situation where currently there are some 24 Water Boards left in The Netherlands.

Deterioration of the quality of surface waters was, among others, apparent in the River Rhine in the 20th century; the load of inorganic and organic waste increased significantly in the 1950s and 1960s[47]. Pollution of the river was at its worst in 1971, when the water lacked oxygen in the downstream sections and aquatic life began to disappear. This critical situation urged the riparian countries of the Rhine to take action, leading to the establishment of the International Commission for the Protection of the Rhine against Pollution in 1950[48]. Combined actions in the riparian countries led to improved water quality in the Rhine after 1971[49] (Fig. 1.2).

The developments in water management also become clear when looking at the consecutive policy documents and water related Acts. The first National Policy Document on Water Management[50] was published in 1968. Water quantity management was the major issue in this policy but some water quality issues were also addressed. In 1970, the Pollution of Surface Water Act[51] came into force. The Act provided for planning instruments, a system of compulsory permits and a scale of charges meeting the full costs of the water quality program. The main focus was on reducing emissions of organic material. Wastewater treatment plants were built, both by large companies to treat industrial effluent and by water boards to treat the wastewater of households and small businesses. Once major reductions in organic wastes had been achieved, the focus switched to heavy metals and organic micropollutants from industry; in 1983, the Pollution of Surface Water Act was included in the more generic "Decision concerning Quality Targets and Monitoring of Surface Waters"[52].

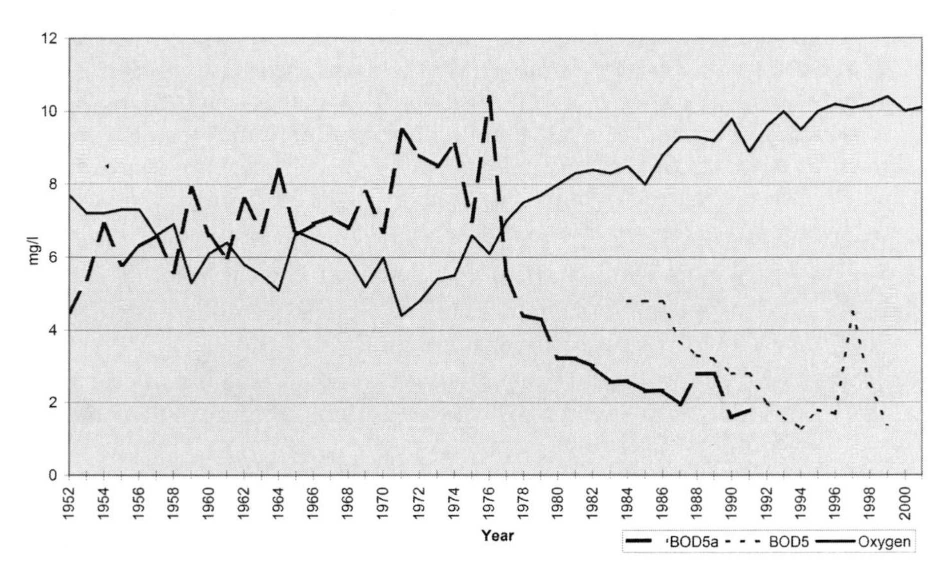

Figure 1.2 Mean annual values of oxygen concentration and BOD (Biological Oxygen Demand) at Lobith (BOD5a is oxygen demand over 5 days with allylthio-ureum, BOD5a measurements stopped after 1992, BOD5 measurements stopped after 1999)

The second National Policy Document on Water Management that was published in 1984[53] paid substantial attention to water quality issues. In the meantime, between 1975 and 1979, the European Union laid out legislation on surface waters[54-57] describing lists of parameters to be monitored, including analysis methods and frequencies[45, 58].

The third National Policy Document on Water Management, published in 1990, had a strong focus on water quality and contained a list of water quality standards. The policy was extended to include emissions from various sources, for instance agriculture (phosphates, nitrates and pesticides), households (metals from gutters and water-pipes) and shipping (tar, oil and micropollutants)[45, 59].

In the mid-1990s, the water quality had improved significantly and political attention towards water quality decreased. This is reflected in the fourth National Policy Document on Water Management, published in 1998[60], that maintained the existing water quality policy but changed the focus towards socio-economic issues. The current water policy is described in the National Water Plan, published in 2009[61], that describes water management within its societal context.

1.1.5 Stages in Dutch Water Management

Water management in Europe has, according to van Ast[62], long been a matter of mere water quantity management. Drinking water supply systems and sewerage systems were already developed and managed in Roman times. Over time, waterways and harbors were constructed and some 1200 years ago, among others in the river Rhine delta area which is currently known as The Netherlands, large-scale hydrological interventions started. The development in water management in The Netherlands from the Roman times until the present day is described in five different stages where the stages 3, 4 and 5 only appeared in the last four decades (also see Box 1.6):

1. Safety-management with emphasis on flood control;
2. Quantity management, not only dealing with keeping the water out, but also to reclaim land from the sea;
3. Sectoral water management: water management focusing on individual sectors like navigation, agriculture, industry, drinking water, etc.;
4. Integrated water management based on water systems, which implies to account for the ability of the total water system to supply all the sectors in a holistic, comprehensive manner;
5. Interactive water management that requires: 1) interaction between the water manager and the factors of the total water system and 2) interaction between the water manager and the different actors in society. This type of water management has four basic components: 1) the water system approach as a starting point, 2) the river basin concept as a policy object, 3) interactive management of water system and people, and 4) sustainable development as the ultimate goal.

Box 1.6 Stages in water management

Similar stages are described by Allan[63], who identifies five water management paradigms:

1. The pre-modern paradigm with limited technical or organizational capacity before 1900;
2. The industrial modernity paradigm with emphasis on engineering until the 1980';
3. The environmental awareness paradigm of the green movement until the 1990's;
4. The economic value of water paradigm until 2000; and
5. The current paradigm with emphasis on political and institutional aspects of water management next to economic and environmental aspects.

By comparing the subsequent Dutch National Policy Documents it becomes clear that there has been a fundamental shift in water management from technocratic water engineering to integrated and participatory water management[64]. Water management in The Netherlands was up to 1960 mostly concerned with discharging the surplus of water into the sea, although this could eventually lead to water shortage in dry summer periods. This policy was reflected in the first National Policy Document on Water Management[50], published in 1968, where water quantity was the central issue (stages 1 and 2). During the 1970s, attention in water management turned to the utilization of fresh water. Water storage and water transport, effective use of fresh water for agriculture, shipping, drinking water and cooling purposes were the main topics in the second National Policy Document on Water Management[53], published in 1984. The policy document added links to economy, for instance the monetary value of fresh water, and brought about attention to water-quality aspects (stage 3).

Because of intensive industrialization and agriculture, water quality management became ever more important. Solutions for the pollution had to be found within a broad context of water management. Therefore, from the late 1980s onward, water management changed to integrated water management (stage 4), interrelating different aspects of the water system like physical planning, ecology and emissions. This was laid down in the third National Policy Document on Water Management[65] in 1989 where water is portrayed as a coherent system and ecology is valued next to the economy[66-68].

Water management became more complex because of the growing inter-linkages between water use and land use, especially in The Netherlands as there is a struggle for space in this densely populated country. A more vigorous coordination between policies on especially the environment, physical planning and water management was needed. This has led to the idea of integrating not only within the water environment, but to integrate the various water management functions into an interactive framework of ecology (including ecosystems, natural resources and biophysical processes), sociology (including consumption patterns, demographics and culture), and economy (including capital, production and labor)[69-73]. The integration should account for the various interests involved including flood protection, agriculture, ecology, public water supplies, transport, recreation and the fishing industry. The scope thus broadens from elements that are linked to processes that interact. Many of these developments are reflected in the fourth National Policy Document on Water Management[74] that broadens the water management policy field to include, among others, physical planning and the socio-economic significance of water, not only as production factor but also as intrinsic feature in the landscape, culture and history of the country.

The fourth National Policy Document on Water Management was the result of a process of public consultation involving a large number of stakeholders. National and regional meetings were held and over 3000 people attended. The public at large was also given the opportunity to submit written reactions on the proposals for the new policy document and some 100 organizations, individuals and institutions responded. Dutch water management in this way adopted the integrated participatory approach (stage 5).

In the year 2000, a Dutch commission studied the necessary adjustments in water management in The Netherlands, taking into account the climate change, sea level rise and land subsidence. It was concluded that water management based on the technical control of the water as performed in the 20th century was no longer valid. The future developments, when longer periods of drought will alternate with more intensive rainfall, require the installation of 'buffer-zones' for the retaining of water for dry periods and for storage of water in wet periods. The social and economic costs of these developments are thought to be less than the costs if no action is undertaken[75]. The concept of sustainability in Dutch water management is translated into the principle of basing measures on natural processes and restoration of resilience of water systems through this approach. Water conservation and buffering in this concept implies to make areas more self-sufficient and less influenced by adjacent areas. Restoration of original dynamic processes will increase the self-regulatory capacity of water systems. The solutions that are developed are often rather technological but involvement of local communities has resulted in, among others, innovative design of water retention areas[76].

1.1.6 Developments in Dutch Water Monitoring

The initial water quality monitoring network was set up to address the evident problems in the river Rhine like the deteriorating oxygen situation. The network design that has developed over the years; the exact monitoring locations, choice of parameters, and sampling frequencies, is determined each year and put down in a monitoring schedule. Such schedules are usually based on the schedule of the preceding years and include only minor changes. However, over time the changes with regard to the original plan can become extensive. Regular evaluation of the network is therefore necessary to determine if the network still satisfies the information needs and to conclude if the monitoring strategy is still valid. Substantial optimization and evaluation studies of the water quality monitoring network were done in 1965, 1978-1981, 1991-1992, and 1996[58]. These studies focused on statistical optimization of the network while the choice of parameters was largely based on legislation. This is a technical/scientific basis combined with a regulatory/standard-driven perspective (see Box 1.10)[58]. Little attention was paid to specification of information needs of decision makers.

After the year 2000, the monitoring network was tuned to the requirements of the European Union Water Framework Directive (WFD)[77]. The WFD in its annex V prescribes in a detailed way the information requirements for water management. These requirements are commonly considered stand-alone tasks without linking them to the water management issues. Where the goal is to provide information to support river basin management, little consideration on the actual water management needs is left in the design of the networks.

The number and distribution of monitoring locations in the Dutch monitoring network shows a large variation over time (Fig. 1.3). The influence of the optimization and evaluation studies is apparent. Between 1952 and 1958, monitoring was performed at four locations in the Dutch part of the Rhine river basin. After extending the network to cover more locations in the Rhine basin as well as locations along the river Meuse and in the delta of both rivers, in 1965 the monitoring network was extended to cover the whole network of the Dutch major rivers. The number of monitoring locations gradually extended to reach its maximum of 224 locations in 1978. The evaluation study that was

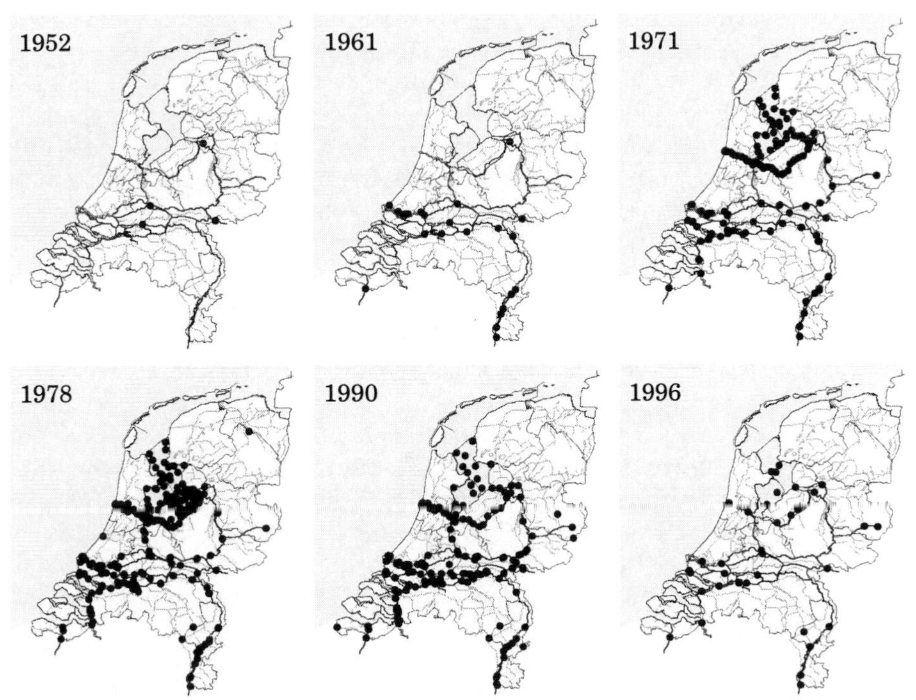

Figure 1.3 Changes in monitoring locations in the network[78]

finalized in 1981, reduced the monitoring network to 149 locations, based on the conclusion that a monitoring network with few locations and a high sampling frequency would yield more information than a network with many locations and a low frequency[79]. The 1992 evaluation study showed that effective information could be derived from concentrating monitoring efforts on few locations and the network was reduced to some 36 locations (Fig. 1.3).

The number of parameters included in the Dutch monitoring network has shown a significant increase over time, starting from 13 parameters in 1952 to over 285 in 1997 (Fig. 1.4). Starting with general parameters like temperature and suspended solids, chloride and pH, the network extended to include eutrophication, metals, organic pollution parameters and radioactivity parameters between 1957 and 1971, pesticides in 1972, and other organic micro pollutants (some 70 different micro pollutants in 1995)[58].

The sampling frequency in 1965 on the major monitoring locations, where the rivers enter (Lobith and Eijsden) and leave (Nieuwe Waterweg) the country, was once a week. On the other locations, the sampling frequency was once in two weeks[80]. After critical evaluation of the data in 1965, the frequency of once a week on the major monitoring locations was lowered to once in two weeks in 1966[81], the sampling frequency for many other locations remained at once in two weeks[82].

In Fig. 1.5, the average number of measurements per parameter for each parameter group is shown. The actual decrease of average frequencies based on the optimization study in 1966 only becomes visible after 1967. The decrease in number of metal samples is striking. It is clear that the parameters with complicated and therefore expensive analytical methods (organic micro pollutants, pesticides, PCB's) are only measured at low frequencies of once per month or once in two months. From the figure it becomes clear that the average monitoring effort for each parameter group is reduced over time. The optimization studies in 1967, 1982 and 1985 are clearly visible as a decrease in most of the parameter groups. The frequencies in the period 1985 – 1992 appeared to be too low for reliable results, and accordingly the 1993 study led to a small increase of the frequency for most parameters.

The number of locations was reduced to almost 1/3 of its size in 1993 as compared to 1992 on the basis of statistical analysis and optimization. Nevertheless, the overall effort, expressed as the total number of measurements, did not change because of the increase in sampling frequency and number of parameters. From 1982 onward, the efforts to run the monitoring network have hardly increased since 2002 (Fig. 1.6) when several new parameters were added, mainly organic micropollutants and pesticides.

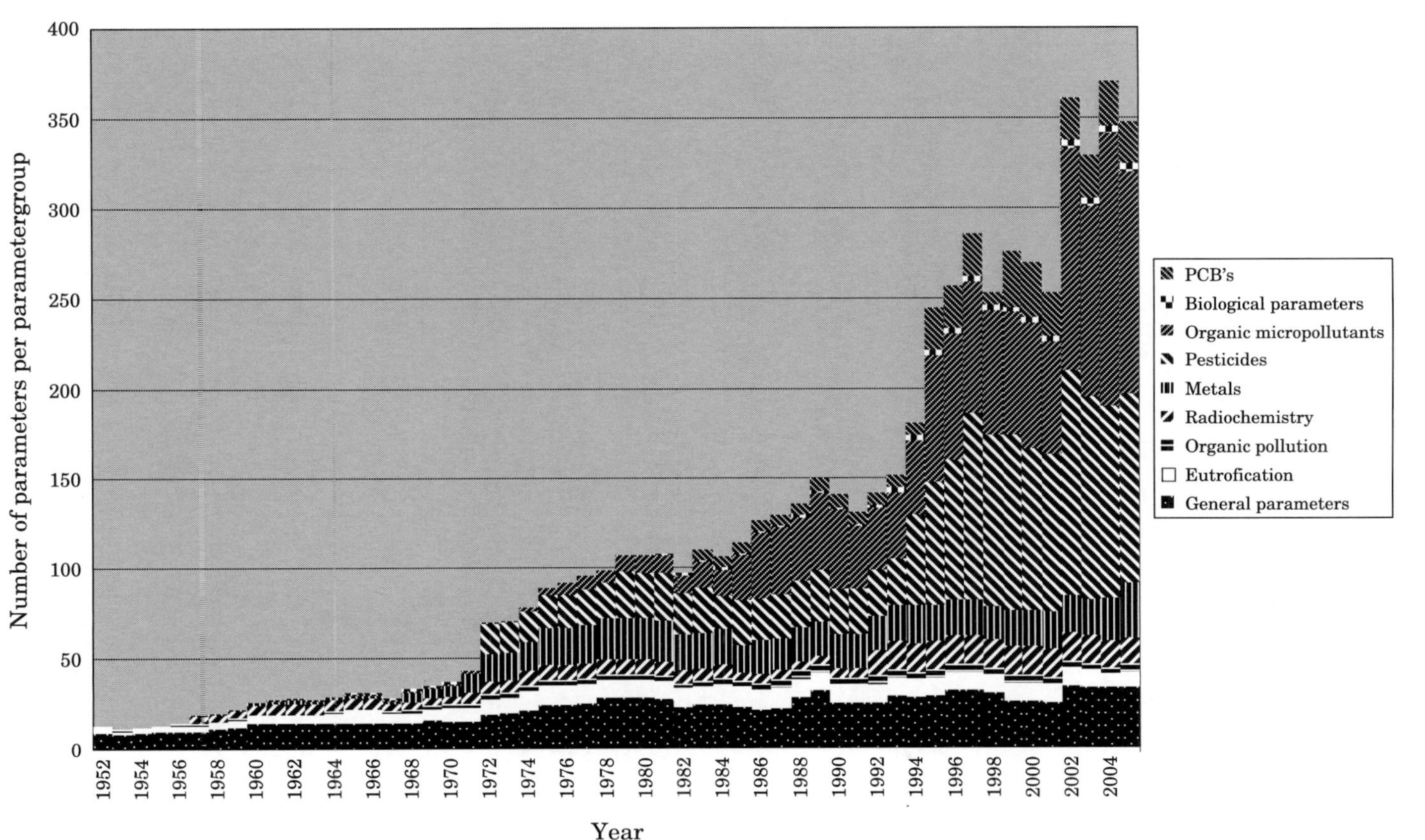

Figure 1.4 Number of parameters per parameter group included each year in the national monitoring network[78]

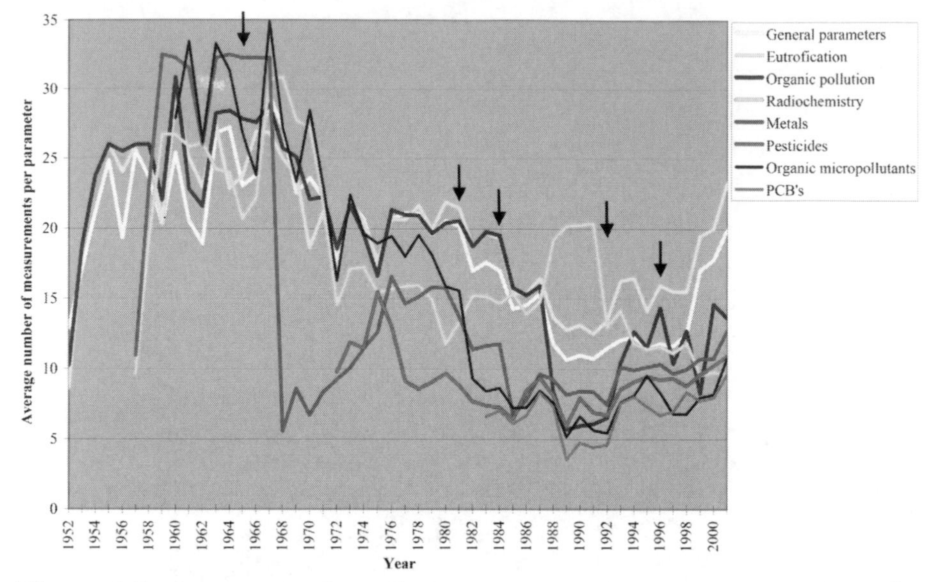

Figure 1.5 Average number of measurements per parameter per year for each parameter group (arrows indicate the years that evaluation studies were performed)

Color image of this figure appears in the color plate section at the end of the book.

1.1.7 The Need for Meta-information

Monitoring networks are not as invariable as they are supposed or thought to be. Year after year, small changes in the networks are implemented. As these small changes are usually scarcely documented, the actual information coming from a monitoring network lies largely in the hands of the monitoring people. Monitoring networks tend to expand over time mainly through the addition of new parameters. This expansion is driven by two developments: 1) new problems arise or are recognized that lead to new parameters to be measured, and 2) new technologies and analytical methods enable measuring of other parameters and handling of larger sets of data. The first point is often research-related and is mostly not policy-driven. The second point is usually of a technical nature without any policy involvement. Together, the expansion of monitoring networks can be largely attributed to information producers. Regular evaluation is needed to limit the growth of the networks or even reduce the monitoring efforts to keep the network attuned to the existing policy goals and the needs of the information users.

Contradictory to this notion, the parameters as selected in the Dutch water quality monitoring network can almost totally be ascribed to existing legislation[58]. The selection of locations and frequencies is done on the basis of hydrological and monitoring expertise and can only be statistically optimized after a period of collecting data.

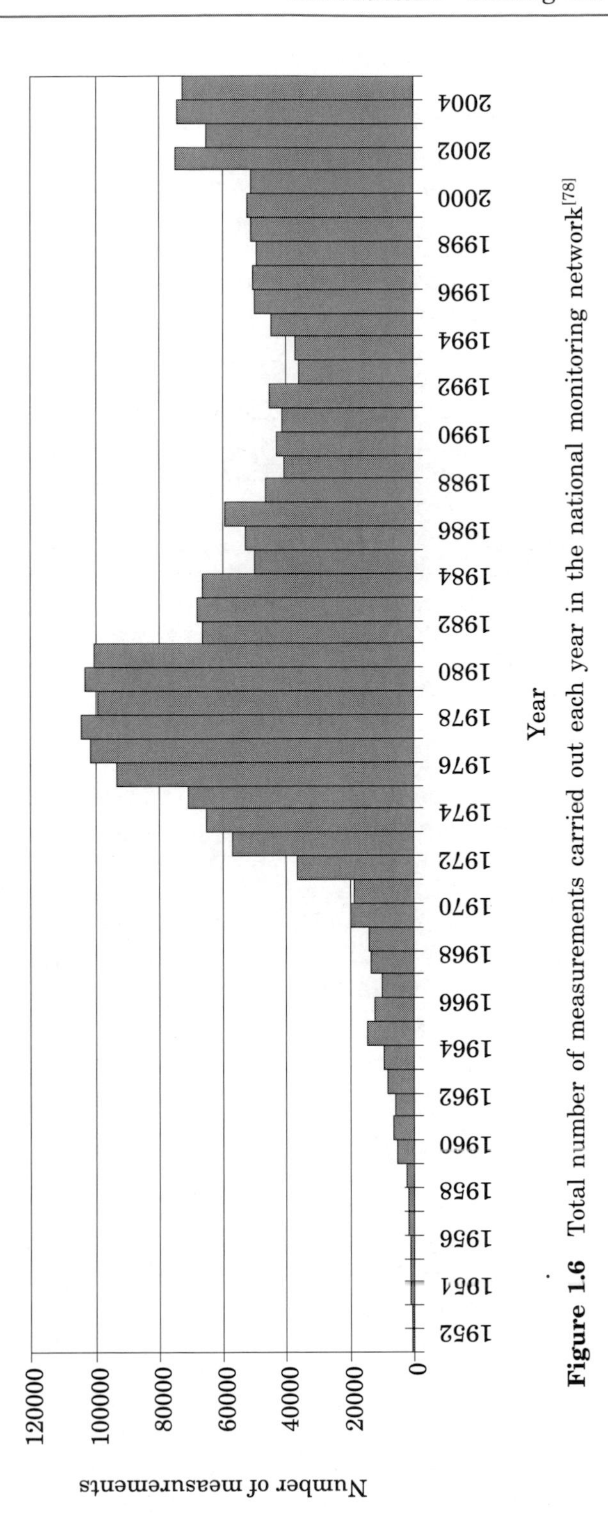

Figure 1.6 Total number of measurements carried out each year in the national monitoring network[78]

Analyzing meta-information on the changes in a monitoring network as done in Section 1.1.6 is helpful to obtain an overview of the developments of the network but also gives insight into the developments of various policy-fields[83]. The developments in the Dutch national water quality monitoring network show that there has been a long process of finding out what information was needed and how this information could best be produced. Initially there were few, rather severe water management problems. In that situation it was possible for decision makers to keep track of the developments with the use of a limited number of parameters. Policies and legislation developed in that period reflect this situation by, for instance, listing parameters and attaching standards to them. Meanwhile, with increasing experience and availability of data, the information could be produced more efficiently. With the increasing complexity of water management problems this parameter-oriented approach showed its limitations with ever-growing lists of parameters. Latest developments in policies and legislation show a trend towards less attention for parameters and standards but there is no new system to replace it yet. The EU Water Framework Directive combines prescribing a list of parameters with standards attached to a more problem-oriented approach where parameters that are produced or used in significant quantities should be included in the network, as described below.

The effort put into specific groups of parameters shows a tendency of increasing attention followed by a 'steady state' and a subsequent decline after the situation for that group comes under control. It appears that in monitoring as in policy, a cycle exists where decreasing interest and efforts follows growing interest and effort. This cycle lags behind the policy life-cycle in time. Monitoring as a consequence has a longer time-span of attention than the time-span of political attention. When including new (sets of) parameters in a network there is a phase of testing; the behavior of the different parameters in the environment is not known beforehand. With the use of time series of data, over time, the behavior in relation to the monitoring network is determined and only then the monitoring of that parameter can be optimized. Then, with decreasing importance, the monitoring effort can decrease. This also explains why monitoring requires long-term planning.

The EU Water Framework Directive has set a Europe-wide standard for water quality monitoring. It prescribes how member states should perform their monitoring. Item 36 of the preamble of the Directive stresses that monitoring should be done on a systematic and comparable basis while Item 42 calls for a committee procedure to ensure such comparability. This requires a well-established monitoring program, which can be developed with the use of the information cycle that is discussed in Chapter 2 of this book. Article 8 states that member states shall ensure programs for monitoring in order to establish coherent and comprehensive overview of the water status. Monitoring should provide for the evidence that environmental objectives as identified by the member states will be achieved (art. 11, 15) or evidence for environmental contamination (art 16). Annex V of the directive sets out the monitoring program essentials[84].

Although it appears that this annex V can replace a regular evaluation of the monitoring network, the contrary holds true; networks are implemented without much consideration if they will provide the necessary water management information. This can hinder proper reporting about the water management efforts, which is also a requirement of the WFD. The WFD leaves a high level of freedom for the member states to implement their monitoring despite all the details provided. The monitoring needs to be tailored to the specific situation in a body of water. The WFD therefore does not make a critical evaluation of information needs redundant.

1.2 THE WATER INFORMATION GAP

1.2.1 The Information Frustration

A large amount of information is collected by water management organizations every year to support the evaluation and development of water management and water policy. Scientists, as information producers, put a great deal of effort in developing and optimizing monitoring networks to collect meaningful and scientifically sound information in a cost-effective way, and in conveying this information to decision makers. Information from monitoring networks is widely used and applied. The networks produce large amounts of data that is overall useful for all users together—individual users may however not always get exactly the information they want. Decision makers, as information users, are informed about the situation in the water system through well-documented reports and presentations[83].

Despite the unprecedented amount of information that is currently available to decision makers, they are not satisfied with the information they get, the criticism focusing on scientists not providing the right information and on providing too much information[85-88]. This has led to frustration from the information producer's side because their efforts appear not to be appreciated. On the other hand, the information users feel frustrated because they seldom get exactly the information they want to have. This dissatisfaction is generally acknowledged in literature[10, 89]. This is the situation that I call the water information gap[83].

1.2.2 How the Gap Developed?

The gap between information users and information producers slowly developed starting in the 1950s when water monitoring as a scientific discipline started. Over the years, with increasing availability of monitoring data, scientists were able, through statistical methods, to make their monitoring networks more efficient by reducing frequencies and number of sampling locations without losing much of the information. Meanwhile, water management faced a growing complexity and also the organizational complexity increased.

In the 1980s it became apparent that the information resulting from monitoring did not always satisfy the policy needs. The attitude towards monitoring became more critical and the criticism focused on scientists not providing the right information (see Box 1.7) and providing too much information[85, 90-93]. Ward and others[92] called this the 'data-rich-but-information-poor' syndrome. They emphasize that water quality monitoring—thereby limiting water monitoring to water quality aspects—is conducted routinely over time at fixed sites without properly defining the exact objectives (the 'why') of routine monitoring. Other scholars besides this emphasize the need for multidisciplinary or interdisciplinary information production that supports Integrated Water Resources Management (IWRM) to improve the situation [83, 94-97].

Box 1.7 Criticism on monitoring programs

"The true state of European water resources and aquatic ecosystems is unknown. Monitoring programs are inadequate or non-existent in many Member States, and where they are in place, their results often remain inaccessible to the public. (...) It is astonishing in a rich continent like Europe that the EU is currently unable to accurately indicate the extent of pollution and disruption of its aquatic resources"[98].

Inconclusive discussions on specific information being essential or obsolete took and take place regularly in many organizations. This leads to frustration from the information producer's side because their efforts appear not to be appreciated and from the information users side because they seldom get exactly the information they want. Scientists are in this situation portrayed as hobbyists that need ever more money to build so-called 'data-graveyards'[99] while decision makers are portrayed as people that constantly change their mind and cannot decide upon what they want[83] (Box 1.8).

Box 1.8 The 'value-action gap'

Generally, there is a difference between what people say and what people do. Blake[100] calls this the 'value-action gap'. He among others concludes that this gap cannot be overcome simply by invoking an 'information deficit' model on the situation but that individual attitudes, people's perception of responsibility, and their perception of the practicality of actions are decisive for people's actions.

This mismatch in water monitoring between information users and producers is a specific part of the policy-science gap (Box 1.10). Within the broad context of the use of environmental information, of which water monitoring is only a small part, policy makers from around the world are calling for more 'useful' information[101]. Information is considered useful when it is 1) salient and context-sensitive; responding to the specific information demands, 2) credible; perceived by the users to be accurate, valid and of high quality, and 3) legitimate; the production of information is perceived to be unbiased[101, 102]. *Salience* relates to the relevance of information: is the information needed to help make a decision? *Credibility* addresses the technical quality of information: Is the information scientifically or technically valid or accurate? Is the information based on my standards of scientific plausibility and technical adequacy? *Legitimacy* concerns the fairness of the information process: Does the system seem unbiased in addressing my values, concerns, and questions and those of others I believe should be included in the process? Scientists that produce too much information that is not considered relevant and useful by decision makers may cause this. On the other hand, information users may not be aware of the existence of potentially useful information.

Box 1.9 Six types of rationality

Hollick[103] distinguishes between six types of rationality; technical rationality aiming at a selection of means to achieve a given end, economic rationality aiming at allocating scarce resources between alternative ends in order to achieve the greatest possible total benefit, ecological rationality aiming at conservation and enhancement of the natural environment, social rationality aiming at individual members that have common needs and desires, legal rationality as a framework to stabilize and institutionalize conflicts with the goal of preventing and settling disputes, which might otherwise threaten social stability, and political rationality aiming at maintaining or enhancing the powers of the group or individual making the decision.

Policymaking is generally seen as a social process. In that process, choices are not only rational but are a sum of various stakes that interact or even counteract, and where societal norms and values are as important as interests[35, 104-106]. This is a general issue that policymakers do not use knowledge and information the way that scientists expect them to (Box 1.9). Many scholars describe this so called policy-science gap. The water information gap is a specific part of this policy-science gap that will therefore be discussed in some detail here.

Box 1.10 The policy-science gap

Boogerd and others have analyzed ineffective communication and non-use of knowledge in implementation of the policy to abate desiccation in The

Netherlands starting from the following three theoretical assumptions to describe the policy-science gap[107]:

1. Two-cultures or two-communities debate. This is the assumption of collective 'programming' of the mind, which distinguishes the members of one group or category of people from those of another.
2. Mismatch between the policy making process and the research process. This explanation assumes that policy makers will not use scientific information unless this information yields recommendations that they consider by intuition politically feasible and easy to implement. Scientists, in turn, compete for funds from different agencies that serve conflicting policy interests.
3. Knowledge on policy problems is socially and politically structured. There are multiple perspectives with regard to complex policy problems and it is difficult if not impossible to determine the 'scientific truth'.

The two-communities model is for instance voiced by Burgess and others, who point at the differences in discursive power between those with technical, professional expertise and lay people, the first being the scientists as information producers, the second policy makers and the public at large as information users[108]. Also Falkenmark adheres to this model when she notes that scientists have difficulties in addressing the system as a whole, with all the physical and socio-economic interactions[12]. She argues that reductionist-oriented scientists have concentrated too much on their partial realities and that policymakers have an economic focus because scientists do not educate them in an interdisciplinary way. Looking at information and information needs, the two-communities model is sometimes explained by the argument that management and policy issues tend to be vague and loosely specified. Policy makers, politicians, the public, and other 'information users', tend to pose questions of the form: 'Is this country safe against flooding?' or 'What will be the consequences of dry years for agriculture?'[109]. Experts, researchers, scientists, and other 'information producers' on the other hand, tend to provide answers like: 'The maximum water level is 34.6 m above mean sea level' or 'pH is 7.8'. Many information users find those answers difficult to relate to their questions[110].

The mismatch model between policy and science is assumed by Bradshaw and Borchers who attributed it to the level of (un)certainty in knowledge that is accepted by scientists and policy[111]. The confidence in a scientific finding within the scientific community grows to a certain level while the level of confirmation of this finding is still low. With an increasing level of confirmation, the confidence level also increases. This is the cognition phase, in which scientific information is disseminated and there is vast scientific discussion. At a certain level of confirmation, there will be scientific consensus. The confidence of society lags behind that of the scientific community, and society needs a certain level of confirmation before the confidence starts building up. If this lower limit of confidence is reached, the volition phase is entered with fast growing confidence within society. Depending on the nature of the scientific finding, the shape of the function will vary[111].

The third model stating that knowledge on policy problems is socially and politically structured is further elaborated by Milich and Varady, who identify inconsistencies between policies, plans and practices in transboundary water management settings[112]. They distinguish four conceptual paradigms in international environmental accords that explain for some of the inconsistencies:

1. The technical/scientific paradigm where concrete goals are established, but management is mostly delegated to organizations dominated by scientists and engineers;
2. The regulatory/standard-driven paradigm where international environmental quality accords are moved towards numerical standards and strict regulation of pollution;
3. The closed paradigm where the process of negotiating international agreements has been restricted to high-level professional diplomats; and
4. The top-down paradigm where ratified international agreements supersede domestic laws and arrangements.

These paradigms deal with different dimensions and may be present at the same time at different levels of treating the issue. Each paradigm represents a certain bias, which, according to Milich and Varady[112], may lead to the following setbacks:

- Too little attention for the specific regional situation of the river basin;
- Political and economic imbalances are reflected in the water management situation;
- Implementation of accords is left to the discretion of the parties;
- Little or no public participation resulting in internal friction; and
- The accords may be too focused on one or few issues, like navigation or 'development'.

These paradigms and their disadvantages can be counteracted through inclusion of non-technical perspectives like social and economic consequences into the process to abate the technical/scientific paradigm, promotion of capacity-building, bringing in a better understanding of the natural processes in the regulatory/standard driven paradigm, transparency through open meetings in a closed paradigm situation, and a bottom-up design through public participation when the top-down paradigm is dominant[112].

All three models in the study of Boogerd and others[107] proved to be valid to a certain degree. They concluded that more attention should be paid to the translation of policy problems from higher levels of political authority to the conceptualization at lower management levels to overcome the science policy gap.

1.2.3 Bridging the Gap

To bridge the water information gap, there is a call for less quantity of data and more targeted, interdisciplinary, tailor-made information[91, 113-117] (that is information that is effective—the information product is tailored to the questions—and efficient—the information is provided at a reasonable and affordable price[118]). The main question is how to produce such tailor—made information. Many scholars suggest that a process is needed of determining the water management problem and the information needs related to it. Such a process should be a systematic effort to consider the purpose of data collection ahead of designing/executing a monitoring program. This is reached through a scientifically sound information needs assessment methodologies that involves the actual users of the information[13, 35, 85, 91, 119-128].

This book describes how to perform such a process. The methodology as described in the next chapters enables decision makers and scientists to decide jointly upon the information that must be produced. The outcome of the information needs specification process is a set of information needs that decision makers recognize and appreciate as representing their requirements and from which scientists are able to develop an information producing system. But for further understanding of the fundamentals of the methodology, in this chapter a closer look is taken at policy problems, their nature and the way they can be structured.

1.2.4 The Nature of Policy Problems

Decision making in general often amounts to choosing among competing values like health, wealth, security, peace, justice, equality, and freedom. Therefore decision-making, including environmental decisions, requires balancing several, sometimes contradictory issues. Consequences of decisions in time, effort, costs, etc. have to be evaluated to come to the best decision. Information to support decisions is collected and valued, and a balance is made until finally a decision is ready. This also requires moral reasoning, which in turn is subject to changing values and different cultures. Mirroring this notion, the way in which a society manages water quality is a telling reflection of political, cultural and economic processes within that society[20, 129, 130]. Or, as Dunn[130] phrases it, "Policy problems are partly in the eye of the beholder. Policy problems are unrealized needs, values, or opportunities for improvement that may be pursued through public action".

Most societies around the globe recognize a broad range of values, such as those involving public health, amenity, recreation, and environmental values like ecological integrity[131]. Water management allows society to define and protect the uses it values from any water resource; all management decisions stem from the context of use-oriented goals. An important reality in water management is that no body of water can be all things to all people: uses frequently conflict, and valuation of water quality depends on the social, political, and cultural contexts of those uses. "Environmental planning and decision-making are essentially conflict analyses characterized by socio-political, environmental, and economic value judgments"[132]. As these sometimes compatible and often competing uses are balanced, water management is moving into a more primary position at the beginning of planning processes rather than as a tail-end consequence[20].

The water-policymaking process is discussed extensively in literature and several models are described or used to explain it[6, 9, 17, 133-139]. The focus in this book will be on the complex and often unstructured nature of policy problems and the need to structure them in order to address the problems. Within this framework, the policy life-cycle of Winsemius is used to describe the general steps in policymaking and as the model for the water policy process (see Box 1.18)[140].

1.2.5 Unstructured Policy Problems

From the description of the developments in Dutch water management (see Section 1.1.4) it becomes clear how problems became increasingly complex over time and will continue to be complex in the foreseeable future[10, 131] (Box 1.11). This complex situation that no single perspective or point of view captures the totality of problems is also called 'emergent complexity'[141]. Water management is as a result in literature currently described in terms of complexity; next to the 'tame' problems (policy problems that can be defined rather straightforward) there is now a class of problems termed 'wicked' (policy problems cannot be definitively described)[142] or 'unstructured'[64, 141, 143] (Box 1.12). Another word used is 'persistent' which is defined as having an even higher degree of complexity than wicked problems[64].

Box 1.11 The complexity of environmental problems

The complexity of environmental problems in general is among others confirmed by the Dutch Scientific Council for Government Policy[143] who reported on a study after the long-term developments in environmental policy. The Council states that problems in environmental policy like climate change and biodiversity, dangerous substances, environmental implications of modern biotechnology, increasing use of energy, and new forms of biological contaminations are not yet manageable within the existing policy frames.

Tame problems are not trivial problems, but problems that can be tackled with some degree of confidence. Tame problems are understood to a degree that they can be analyzed using established methods, and is clear when a solution is reached. Wicked problems on the other hand, are problems that are characterized by a high level of technical and strategic uncertainties (there is no objective measure of success), and that have to be solved in a situation in which the parties involved are highly dependent on each other while each party has its own moral, political or professional preferences[144, 145]. This book aims to deal with both tame and wicked problems.

Box 1.12 Post-normal science

Funtowicz and others illustrate the complexity of water management problems in a diagram that plot systems' uncertainties against decision stakes[146]. In this diagram (see Fig. 1.7) they place applied science and professional consultancy at the level where uncertainties and stakes are relatively low and call the science, that has evolved in the present-day unstructured decision situation, post-normal as compared to classical science. Post-normal science is issue-driven science that relates to environmental debates where typically facts are uncertain, values are in dispute, stakes are high, and decisions are urgent. The old approach to knowledge and information in this post-normal situation will no longer provide the desired results. The classical distinction between hard, objective scientific facts and soft, subjective value-judgments is now changing and hard policy decisions are made where the scientific inputs are irremediably soft. The focus shifts towards the quality of the process, which depends on open dialogue between all those affected. Where the quality in 'normal' science was assured by peer review, for quality assurance there is the need to turn to an 'extended peer community' that includes all those with a desire to participate in the resolution of the issue[146].

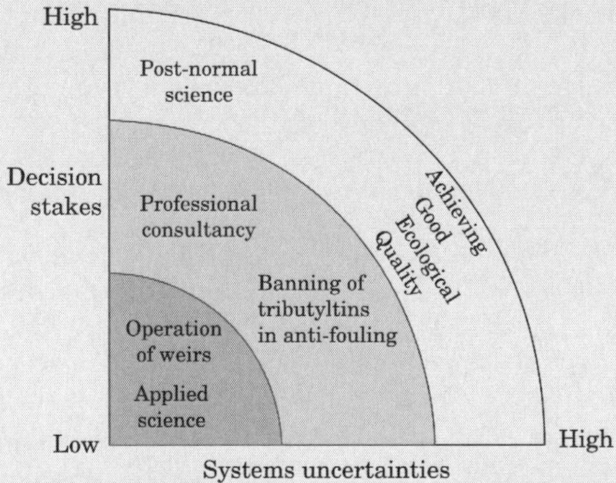

Figure 1.7 Post-normal science in water management (based on[146])

Some water management examples are added to Fig. 1.7 to make all this more tangible. The operation of weirs in rivers to maintain a certain water depth for navigation is an example where the uncertainties are relatively low; reliable hydrological models are present to support decisions here. A wrong decision will in general be easily adapted, as the system is designed to monitor the water level and tune the weirs to it. An example where the uncertainties were higher was the situation where malformations and sex change were found in sea slugs and oysters. There were hints pointing at tributyltins, a substance used as anti-fouling in ships, as the main source. The stakes were high for the navigation sector as alternatives were expensive. Only after extensive studies to diminish the uncertainties, a decision on banning tributyltins was taken[147, 148]. The third example is the goal set in the EU Water Framework Directive for all member states to achieve a Good Ecological Status within 15 years for all the waters, based on the aim to establish the basic principles of sustainable water policy in the European Union[77]. The decision stakes are high as the entire society is involved through industries, navigation, recreation, housing, etc. The systems uncertainties are high, as there is little exact knowledge on what measures should be taken to achieve this situation without bringing harm to the social and economic situation in the countries.

No distinction will be made in this book between the different terms used related to complexity, like persistent or wicked, but they will be referred to as unstructured problems. From this discussion it may be clear that water management has to perform in a complex situation and that water policy and water management problems are unstructured. The aim for sustainability is an important reason for the inherent uncertainty of water policy problems, as especially the future needs are subject to values and therefore highly subjective. It is nevertheless clear that the sustainability goal requires policies to take a long-term perspective. From this it follows that information to support a long-term perspective should include possibilities for future predictions.

1.2.6 Problem Characteristics

Policy problems have several characteristics that are important when looking at the information needed. Dunn[130] described four characteristics of policy problems:

1. Interdependency of policy problems. Policy problems in one area frequently affect policy problems in other areas. One example is the problem caused by high water flows in The Netherlands where water management has to balance the socio-economic effects—like nuisance and damage from flooding—and spatial planning—for instance the desire to build houses in floodplains —next to the more technical issue of safety—like construction of dikes - in its considerations[149]. As a consequence, information to support water management in this situation has to include such items as spatial planning and risk of damage next to flood control information. This characteristic of a policy problem is an important aspect of the complexity of water management issues.

Box 1.13 Mindframes

The mindframe is the window through which people view the world. It is the frame within which someone's thinking takes place, includes the representation of each person's window (view) on the world, and also contains a notion of outer limits of the frame. It is the mental model that acts like a 'filter' through which the problem situation is observed[150]. The word 'frame' also links to the concept of framing environmental issues[151, 152]. Difficulties in communicating find their root in differences in mindframes; people discuss the same world, but each person sees things that others may not see, especially when people from different disciplines are communicating[151] [153]. In addition, Kolkman and others[154] state that an actor's mental model restricts information flows to only those aspects that affect the actor in question denoting that every scientist or decision maker will have blind spots and none can singly encompass the whole system[146, 155, 156].

2. Subjectivity of policy problems. The way a problem is defined depends on the worldview or societal attitude of people. An attitude is not something static but may change over time. Different terms and descriptions are used for the characteristics of such attitudes. Perry and Vanderklein for instance describe an array of worldviews (or mindframes (Box 1.13)) ranging from anthropocentric to 'ecocentric' or 'biotic integrity'[20]. From the anthropocentric point of view, water is an economic good that can be used. Bad use will lead to economic loss. Water management can in this view put restraints on the use of water for instance by setting environmental quality standards. From the biotic integrity point of view at the other end of the scale, barriers have to be put up to abate any deterioration of ecosystems. Approaches focus on management to retain unimpacted conditions or at least to prevent further degradation. The most common approach nowadays is midway along the valuation continuum between the anthropocentric approach and the 'ecocentric' approach and can be characterized as looking for the 'best attainable quality'[20]. These worldviews are part of the respective mindframes and consequently of the rationale behind decision-making (Box 1.14).

Box 1.14 Various worldviews

Regier and Bronson[157] describe four similar worldviews: exploitist, utilist, integrist and inherentist. The exploitist sees little or no unique value in purely natural phenomena as such; only natural phenomena that can be exploited as a resource a value is added. The utilist sees benefits and costs in natural phenomena. The integrist considers humans as part of nature and humans should take care not to cripple natural processes. The inheritist puts high ethical principles as a constraint on human activities. Exploitists will emphasize information on direct impacts, while integrists will stress the importance of contextual historical information.

3. Artificiality of policy problems. A policy problem only exists when someone judges about the (un)desirability of some existing situation. Environmental problems are therefore at their root social problems[158, 159]. The recognition of a policy problem as such links to the subjectivity of problems and is largely related to the characteristics of problems that go through the issue-attention cycle as described by Downs[160] (see below).
4. Dynamics of policy problems. As time goes by, the nature of the problem may change. This may be due to a higher scientific confirmation level, influence of measures taken or because of a changing worldview or mindframe. An example of the latter is the shift of responsibility for industrial accidents from employee to employer in the western world over the last century. A changing nature of the problem will require different information over time.

The consequence of these characteristics is that a policy problem is hard to isolate because of its interdependence with other policy problems and is also hard to define because of its subjectivity and artificiality. Then, when the problem is isolated and defined it may have changed again because of its dynamic character. It is clear from the above that present-day water policy problems are often unstructured. Producing suitable information for policymaking is therefore complicated.

The interdependency of policy problems is recognized in the concept of Integrated Water Resources Management (IWRM). IWRM is generally viewed as the process that is needed to achieve sustainability in water management in a complex policy situation[5, 161]. The water policy and water management process is in this book considered as dealing with social, economic and environmental factors, that integrates over surface water, ground water and the ecosystems through which they flow as well as considers the differing values attached to these issues. A multi-disciplinary or preferably interdisciplinary approach is needed to implement IWRM (Box 1.16). Information to support this approach must consequently be interdisciplinary and be dealt with in an interdisciplinary way[162, 163]. In this book, IWRM is considered the generic concept of contemporary water management[164] (Box 1.15).

Box 1.15 IWRM and AM

Integrated Water Resources Management (IWRM) is defined in many different ways. GWP for instance defines IWRM as a process that promotes the coordinated development and management of water, land and related resources in order to maximize the resultant economic and social welfare in an equitable manner without compromising the sustainability of vital ecosystems[5, 135]. Jaspers[161] defines IWRM as the management of surface and subsurface water in a qualitative, quantitative and environmental sense from a multi-disciplinary and participatory perspective. There is a focus on the needs and requirements of society at large as well as environmental needs with regard to water at the present and in the future in both definitions, thus aiming at maximum sustainability in all senses.

There is emerging insight that the perspective of IWRM is too narrow as the ability to predict future drivers, as well as system behavior and responses, is inherently limited[146]. The current mainstream is that IWRM should therefore be based on an Adaptive Management (AM) approach, which is defined as a systematic process for continually improving management policies and practices by learning from the outcomes of implemented management strategies[165, 166]. This book is based on the concept of IWRM, but as the methodology developed in this book is based on a continuous learning process, it also links to the concept of AM[167].

The subjectivity of policy problems finds its root in the fact that policy problems are socio-political constructs, and defining and structuring of a problem is always a matter of choice, implicitly or explicitly. Hisschemöller and Hoppe distinguish between four different classes of policy problems through a typology of these problems based upon two dimensions; the level of certainty about the relevant knowledge, and the level of consensus on the relevant norms and values. A policy problem is unstructured if there is no consensus on the norms and values nor on the relevant knowledge. The unstructured problem can become moderately structured if consensus is reached on the norms and values—'ends'; if the outcome is more or less certain. The problem can become illstructured if consensus is reached on the relevant knowledge—'means'; if the number of alternatives is limited. Following Hisschemöller and others[168] to better distinguish between moderately structured (means) problems and moderately structured (ends) problems, I will refer to the first as ill-structured and use the term moderately structured for the second. Note that the term ill-structured is also used in literature to indicate problems that in this book are called unstructured[64]. A policy problem finally is structured if consensus is reached on the norms and values as well as on the relevant knowledge[169]. In another view on this quadrant the term 'wicked problem' is applied to the situation where the consensus on the relevant norms and values, and the certainty about the relevant knowledge is low. Ill-structured problems are in this view termed political problems while moderately structured problems are termed scientific problems[145]. This view is not adopted in the context of this book because not all water management problems are wicked problems and specification of information needs in these situations still requires problem structuring. The typology is particularly relevant as the methodology for specification of information needs will concentrate on developing certainty about the relevant knowledge. Hisschemöller and others conclude that the role of scientists in the four problem situations differs where each role requires different types of information[168]:

1. In a structured problem situation there is consensus about the knowledge needed and the values at stake. Scientists can play the role of problem solver in this situation, by providing the necessary knowledge.
2. In a moderately structured problem situation, there is consensus about the values and policy goals but the means to achieve those goals are under discussion. In this situation, scientists can advocate specific viewpoints.
3. In an ill-structured problem situation there is discussion about the values while there is agreement on the knowledge needed. In this situation, scientists can act as mediator.
4. In an unstructured problem situation scientists have the important task of identifying and clarifying the problem.

Box 1.16 Transdisciplinary science

For scientists, the practical challenge of integrated assessment in water management may be best referred to as to make a shift from multi-disciplinary science to interdisciplinary science, where multi-disciplinary implies studying an issue starting from different mono-disciplinary angles while interdisciplinary implies studying an issue starting from the problem that has to be addressed[170]. Interdisciplinary science can be taken a step further by including non-formal knowledge (for instance knowledge derived from experience) and knowledge that is not connected to a specific scientific discipline. This is called transdisciplinary science[171].

Problem situations may be classified as one of these four problem types. But as problem situations may shift in time they may change to another type over time. The policy life-cycle as described by Winsemius[140] or the five phases in the policy-making process as described by Dunn[130] in essence describe the process in which an often unstructured problem situation is turned into a structured one by adding goals and the related relevant knowledge to the problem situation. Only after the problem situation has become structured, the problem comes in the 'control' phase. From the policy life-cycle[140], the phases in the policy-making process[130] or the idea that a certain confidence level is needed before society or policy makers acknowledge the problem[111] one could argue that really unstructured problems do not exist—there has to be a particular level of certainty about the relevant knowledge. The pitfall in this is that the decision maker through his/her mindframe is blinkered or biased in his/her perception of the problem situation and from this perception may approach the wrong problem. Policy makers show the inclination to move away from unstructured problems to more structured ones even at the cost of losing touch with the true complexity and normative volatility of the problems as experienced by others. They show a marked tendency to ignore information that may complicate the policy problem under scrutiny. Although this is not necessarily done deliberately, they run the risk of tackling what is called 'errors of the third kind', that is solving the wrong problem[130, 168, 169, 172, 173]. To avoid this risk, ample attention is needed for problem structuring. Structuring or defining the problem is of major importance. Searching for information needs usually involves dealing with uncertainty about the relevant knowledge and therefore with unstructured or moderately structured problems.

To illustrate the typology as described above, a quick look will be taken at the development of water quality issues. Water quality was a structured issue until the 1970s. Pollution originated from point sources; nutrients largely came from urban wastewater and a variety of other pollutants could be ascribed to certain (types of) industries. Policies could be developed to reduce the pollution from these sources, like development of waste water treatment and licensing of discharges. With reduced pollution levels in surface waters, the situation became more complicated; the pollution problem was no longer attributed to few sources but there was now a range of sources for many pollutants, either diffuse or point sources. Costs to further reduce emissions from point sources may become higher than costs for taking measures to reduce emissions from diffuse sources. The uncertainty has increased and the costs to achieve results may become very high. The problem situation as a consequence became unstructured (Box 1.17).

Box 1.17 Different types of uncertainty

Koppejan and Klijn[174] distinguish three types of uncertainty in dealing with unstructured problems that are directly related to subjectivity: substantive, strategic, and institutional uncertainty. *Substantive uncertainty* is the uncertainty about the content of the problem. It relates to the information and knowledge on the nature of the social problem and how it can be solved. The interpretation of the meaning of information adds to the substantive uncertainty. *Substantive uncertainty* links to the dimension of certainty about the relevant knowledge that Hisschemöller and Hoppe[169] identified. *Strategic uncertainty* relates to the perceptions and objectives of the actors involved, but also on their response and anticipation on each other's strategic moves. Institutional uncertainty originates in the interaction between the institutional backgrounds of the parties involved. The behavior of the actors is guided by the tasks, opinions, rules and language of their own organization. Strategic and institutional uncertainty together are linked to the dimension of consensus on relevant norms and values that Hisschemöller and Hoppe describe, but Koppejan and Klijn include the influence of institutional settings as well as the strategic behavior in the process towards reaching consensus. This influences the way the process should be organized with which the problem has to be structured.

Artificiality of policy problems relates to the fact that a problem must be recognized as such. This characteristic basically links between the characteristics of subjectivity and dynamics of policy problems; there must be an urgency (subjective) to table the issue as a policy problem and the issue can disappear again over time due to the dynamic character of policy problems. But who can initiate an issue as a policy problem? The policymaking process according to Dunn[130] is a series of interdependent phases arranged in time. The issues on the political agenda are institutionalized. Dunn distinguishes the next five phases:

1. Agenda setting, in which elected and appointed officials place problems on the public agenda;
2. Policy formulation, in which officials formulate alternative policies to deal with a problem;
3. Policy adoption, in which a policy alternative is adopted with the support of a legislative majority, consensus among agency directors, or a court decision;
4. Policy implementation, in which administrative units, which mobilize financial and human resources to comply with the policy, carry out an adopted policy; and
5. Policy assessment, in which auditing and accounting units in government determine whether executive agencies, legislatures, and courts are in compliance with statutory requirements of a policy and achieving its objectives.

Hisschemöller and Hoppe state that those actors who have the power to decide on the policy agenda also have the power to choose the problems they like to solve[169]. Agenda setting is in this view a matter of power. Downs identifies a larger group that can stage a problem in what he calls the issue-attention cycle[160]. Downs distinguishes five stages in the issue-attention cycle. The first stage is the 'pre-problem stage' (stage 1), in which a specific problematic issue is identified by experts or interest groups. This implies that any stakeholder can put an issue on the political agenda but needs a certain influence (power) to realize this. For instance, the book "Silent spring" by Rachel Carson in 1962 and the report "Limits to growth" by the Club of Rome in 1972 raised the public's attention to the gravity of environmental issues, entering them on the political agenda. Unfolding events, like for instance severe floodings, make the public become aware of the issue in the 'alarmed discovery and euphoric enthusiasm stage' (stage 2). The ability of society to deal with the issue is discussed. A general assumption is often that a (technical) solution is at hand. Politicians and the public become aware of the (sometimes monetary but often societal) costs involved in dealing with the concern during the 'realizing the cost of significant progress stage' (stage 3). When either the issue is discouraging or solutions to the issue are found, interest is weakening in the 'gradual decline of public interest stage' (stage 4). Often by this time a new issue enters stage 2, distracting the public attention. In the 'post-problem stage' (stage 5), there is limited attention that may at times flare up for a short period. Institutions, programs, policies and legal acts created in response to the issue remain as legacies. The issue (or aspects of it) may capture public interest for a second time and go through the cycle again[160, 175]. In information production, this latter aspect is less relevant as information is only needed after an issue is tabled.

The dynamic nature of policy problems is portrayed in the different phases that literature attaches to the policymaking process. This is, for instance, described in the model of the policy life-cycle as described by Winsemius[140] (Box 1.18).

Box 1.18 The policy life-cycle

The policy life-cycle[140] builds on the issue-attention cycle of Downs[160] and links to the phases in the policymaking process as, for instance, described by Dunn[130]. Downs, Dunn and Winsemius start off from a problem situation and describe several phases. Winsemius[140] describes the policy phases from a politician's point of view, putting the political importance at the center of his description. The political importance in this framework follows the stages of interest of Downs. Dunn, as a political scientist, describes the phases in terms of government's actions. Policy adoption is an essential step in his view.

The policy life-cycle describes four phases in a cycle of policy making, ranging from the moment that an issue enters the policy domain until it is under control[140] (see Fig. 1.8):

Problem recognition: there is discussion in this first phase if there is a real policy problem. Signals from society (for instance from scientists or NGO's) are evaluated and recognized by policy as a policy problem. The political attention, and with this the political importance, is rapidly growing in this phase. There is nevertheless a certain level of discord about the importance of the problem between different parties involved. The second phase is entered if the problem is generally recognized as a policy problem.

Policy formulation: during this phase, consensus is built on the size and nature of the problem at hand and a policy to solve the problem is formulated. The societal confidence level, as described by Bradshaw and Borchers[111] is rapidly building up. The political importance of and interest for the problem reaches its highest point by the end of this second phase. The level of discord is relatively small at the end of this phase, which is in line with a reduced science—policy gap because of the higher level of confidence about the problem. When the policy formulation is done and policy measures are identified, the problem enters the third phase.

Policy implementation: The third phase is the actual execution of the formulated policy; the policy measures become operational. The problem situation is now dealt with by the responsible organizations/institutions and political attention is rapidly decreasing. Usually various other issues are trampling to gain political attention in the Problem formulation phase and as soon as policies enter the implementation phase, political attention swiftly shifts to other issues (this is the main reason that the policy arena is characterized by very short attention spans[176]). After some time, the measures take effect and the problem comes under control; the fourth phase.

Control: When the problem is under control, the emphasis is on securing the situation and improving the conditions. The political importance of the problem now reaches a very low and slowly diminishing level of attention. In case the policy measures do not lead to the expected effect, one of the previous phases is entered again.

Winsemius describes that there is a level of discord in the recognition phase that decreases over time towards the end of the phase of policy formulation. This level of discord concerns the uncertainty about the nature and extent of the problem and its causes and effects, and finds it basis in the science-policy gap. In this light it is interesting to see that Scheffer and others[177] suggest that the political attention can deliberately contribute to the science-policy gap as noted by Bradshaw and Borchers[111] where they describe the response of societies to new problems. This societal response depends on 1) slow detection of new problems, 2) time lag in switching from ignoring to recognition of a problem, 3) downplay of the severity of the problem by a credible party, 4) absence of a decision-making authority, and 5) powerful stakeholders who benefit from delaying regulation. Note that the policy cycle is present on various management levels and that not each level may be in the same phase of the cycle[20].

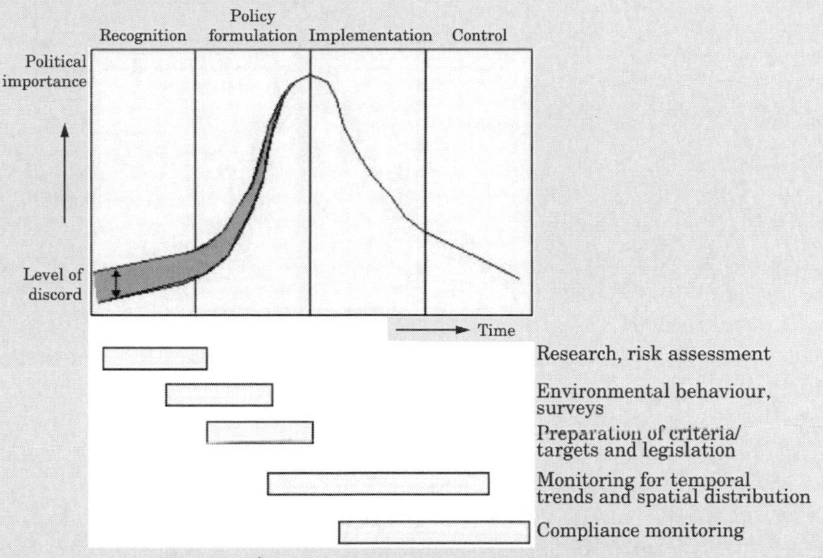

Figure 1.8 The policy life-cycle[140, 178, 179]

Cofino[178] linked information on water management aspects to the policy life-cycle (Fig. 7). In the phases problem recognition and policy formulation, much attention is given to rapid information on the sources of the problem, the risks attached to it, and the processes in the environment. Research and surveys together with information on environmental behavior may identify the causes and effects of environmental problems. These efforts serve to increase the level of confidence and reduction of the level of discord in the problem recognition phase, and to assist in formulating policy measures in the policy formulation phase. In the course of the second phase, as part of the policy formulation, specific criteria and targets are developed to define the level at which the problem can be regarded as 'under control'. The end of the second and during the third phase puts more attention put on longer-term information like trends, especially when measures are taken. Finally in the fourth phase, monitoring is needed to keep track of the problem, but as the problem comes under control, emphasis will be on compliance monitoring. This will be further discussed in the next chapters.

The policy life-cycle grasps the dynamics of a policy problem and makes them better manageable by structuring the problem into four phases that each has their own characteristics. The level of discord links the dynamics of the problem to its subjectivity. In terms of problem structures, the unstructured problem situation in the problem recognition phase changes towards an ill-structured problem situation through input from scientific knowledge. Policy formulation then adds norms and values to the situation, changing the unstructured problem towards a moderately structured problem. In practice, adding scientific knowledge and formulation of policies will iterate until the problem becomes manageable. Then, in the policy implementation and control phases, the relevant knowledge is defined and the problem becomes structured.

1.2.7 Coping with Policy Problems in Information Production

As described in the previous section, the concept of IWRM was adopted to deal with the interdependency of policy problems. The basic criteria for IWRM are, according to the Global Water Partnership (GWP)[5]; 1) economic efficiency in water use, 2) equity, and 3) environmental and ecological sustainability. To reach these criteria, three elements are distinguished: 1) the enabling environment, 2) the institutional roles, and 3) the management instruments. The enabling environment constitutes the policies and legislation that enables all stakeholders to play their respective roles in water management.

More recently, the term 'water governance' was introduced, depicting a change in thinking about the nature of policies. The notion of government as the single decision making authority has been replaced by a more contemporary, multi-scale, polycentric governance. Governance takes into account that a large number of stakeholders in different institutional settings contribute to policy and management of a resource. Governance differs from the old hierarchical model of government in which state authorities exert sovereign control over the people and groups making up civil society[180].

Water management is based on certain (implicit or explicit) principles, rules, and decision-making procedures that enable convergence of stakeholders' expectations. Such a set of principles, rules, and procedures is called a regime. There are five central elements that describe transboundary regimes: policy setting, legal setting, the institutional setting including the actor networks, information management, and financing systems[181]. Water policies that are in place in a country can be found in the formal documents which contain current and future water management strategies. They refer to the goals of government, or other organizations and strategies to reach those goals. The legal framework consists of the full set of national and international laws and agreements. The institutional framework includes formal networks but also informal multilevel actor networks. Information management is needed to assess the current situation and existing vulnerabilities to develop understanding of the possible futures, but also to monitor policy progress. The information produced should reflect multiple perspectives, which entails that governmental and non-governmental stakeholders should be invited to articulate their information needs and formulate the production of information. Sufficient resources should be available to ensure sustainable water management. In particular, climate change adaptation requires significant additional financial resources. Financial as well as ecological sustainability can be improved by recognizing water as an economic good and recovering the costs as much as possible from the users. Cost recovery from the users of the resource is an important funding source which can be directly linked to the intensity of use, but equity considerations need to be taken into account.[182].

Box 1.19 Five reasons for public participation

Van de Kerkhof and Huitema[183] describe five reasons for public participation. First of all, participation may increase public awareness and acceptance of the problems that the water manager faces and of the measures that need to be taken to solve these problems.

Secondly, participation may lead to better decisions as it enriches the decision-making process with relevant viewpoints, interests and information about the water issue that could not have been generated otherwise. It helps to rule out overlooking something, which in turn may improve the decisions.

Thirdly, participation may increase the legitimacy of decision making, as it enables the stakeholders to engage in deliberation about the decisions that need to be taken.

Fourthly, participation may increase the accountability of decision making, as participants get an inside view in the decision-making process and they become co-responsible for the decisions that are made and the actions that are taken.

Finally, participation may result in learning. Stakeholders, government and scientific experts enter into a dialogue and, by interaction and debate, they learn how to collectively manage a river basin and deal with conflicting views and interests.

Box 1.20 The subjectivity of policy problems

The *subjectivity* of policy problems can be divided over substantive, strategic and institutional uncertainties. Koppejan and Klijn provide a range of issues that need to be dealt with to manage these uncertainties and they provide advice on how to implement their recommendations[174]. To deal with *substantive uncertainty*, the objective is not to create consensus but to create an enabling environment for joint image building and cross-frame learning (see for instance Dewulf and others[151] for a discussion on how frame differences can lead to different interpretations of the situation), and to develop interesting and appealing solutions. This should lead to a common ground or 'negotiated knowledge': consensus about insights that are supported by research findings. A precondition to reduce *strategic uncertainty* is the willingness of parties to cooperate. To achieve this, openness and transparency about the process and roles of participants, and availability of information is needed. This can create a platform for the parties to interact. Dealing with *institutional uncertainty* implies that a new institutional design is needed; a new policy paradigm invokes new relationships between parties and institutions. One essential element in the new institutional design is the development of a new set of rules to organize the process. That process must consequently be open, transparent, and participatory.

Turner[184] distinguishes a spectrum of deliberative and inclusionary processes ranging from better information provision and consultation to fully-fledged participatory processes in which stakeholders are a component of the decision-taking mechanisms. The shift towards participation may lead to a social learning process that brings better decision making, better dissemination and transparency, and building of trust and accountability. Another differentiation is made between interest-based participatory processes where one tries to make groups with different interests in the matter (stakeholders) negotiate different strategies and policies in an area, and value-based participatory processes where one appeals more to people as citizens and hopes to achieve a consensus on which values to protect and how[172]. In reality there will always be some mixture of these two.

Participation requires close and regular contact (like through a common physical location and working in teams) between actors as well between different disciplines[174]. It should be realized that although it is often assumed that participation is a process with an outcome as we intend it to be, it may be by definition a paradoxical process that the more planned and anticipated, the less it is there[185].

Participation of stakeholders is, as discussed above, imperative and can be supported by mechanisms like information exchange and capacity building (Box 1.19). The institutional roles in a participatory process have to be clear. This includes the demarcation and matching of responsibilities between actors, co-ordination mechanisms, identification of jurisdictional gaps or overlaps, and authority and capacities for action (mandate). Implementing IWRM becomes difficult, if not impossible, without a proper institutional setting. The instruments to manage the process are the tools and methods that enable and help decision makers to come to rational and informed choices. Monitoring, information systems and communication are identified as important management instruments[5].

The methodology for specification of information needs as described in this book is one of the management instruments. There are two important aspects from the description of IWRM related to this: 1) information is needed to support stakeholders as well as water managers, and 2) this information should be available on economic, social and ecological aspects of water management to enable an interdisciplinary approach. These two are the essential elements to deal with the interdependency of policy problems and necessary for successful implementation of the methodology (Box 1.20). It is therefore necessary, as Koppejan and Klijn[174] stress, that there is willingness among the actors to be closely involved in a process of mutual learning and that this process starts before a decision is made on what the problem actually is. The decision on what the problem is, therewith reducing the substantive uncertainty, is consequently part of the process.

Box 1.21 Dealing with unstructured problem situations

What is needed in an unstructured problem situation is a process of reflection, action, and political strife towards systematically structuring of the problem[168, 169]. Hisschemöller and Hoppe define four conditions to problem structuring (or problem finding). The primary condition for problem structuring to be set in motion is that at least some segments of the official policy elite start interacting with those who have alternative views on the problem. The second condition is that all actors involved are willing to participate. The third condition for problem structuring to be successful is that it should address concrete cases and the experiences of those involved. The fourth condition for successfully applying learning as a policy strategy is that the decision is not taken before problem structuring has produced new insights on the problem and its potential solutions[169].

Mostert and others[186] distinguish a total of 71 factors that foster or hinder social learning in participatory river-basin management that will however not be discussed here. These factors nevertheless confirm the aspects mentioned by Koppejan and Klijn[174] and Hisschemöller and Hoppe[169] and add more detail to them. Among the top 10 key factors that foster social learning are for instance openness of the project and clear expectations. Among the top 10 key factors that hinder social learning are for instance lack of clarity about the role of stakeholder involvement (for instance form, timing, and aims), and lack of stakeholders' belief that their inputs would make a difference.

The methodology for assessing information needs, as described in this book, is expected to be able to deal with uncertainty about the relevant knowledge/information, which is part of the substantive uncertainty. Therefore it needs to be able to deal with unstructured or moderately structured problems (Box 1.21). Scientists, as information producers, must support information users by clarifying the problem and/or provide alternatives of scientific knowledge (see Section 1.2.6). The methodology must act as an aid in this clarification process.

The artificiality of policy problems is not a real point of concern for the information production, as this process mainly affects water management policy and legislation. It does however influence the type of information that is needed. Nonetheless, artificiality is closely related to subjectivity and will therefore not require additional action from the viewpoint of information production.

The dynamic nature of policy problems demands a methodology that is flexible. Policy makers ask for information that responds to their questions. As these questions change over time, as depicted in the policy life-cycle, the information needs change. Information production, as a consequence, must be flexible and the methodology should be able to support this flexibility. On the other hand, expectations about the adaptability of information production should not be too high; information production is usually a process that requires time and policy attention often changes more rapidly than information can be made available.

Problem structuring is imperative for the prevention of errors of the third kind; solving the wrong problem[172]. To this end, the subjectivity of the problem should be reduced. The essential precondition to achieve this is that the parties involved are willing to cooperate. A reticent attitude among relevant parties can be overcome by creating an enabling environment that is appealing for those parties. This, among others, entails addressing situations that link to the experiences of the participants. To attain such a link, a set of clear rules must be available, the process must be transparent, the roles of the participants must be known, and all this must be openly communicated. These are the elements that set the conditions for learning about the different parties' frames and for developing a common understanding of the problem. The common view on or 'negotiated knowledge' of the problem as developed can be considered to be objective, at least to the participants, as it incorporates the collective views.

The interdependency of policy problems has to be addressed next to structuring the problem. As stated, the concept of IWRM challenges the interdependency by promoting an integrated approach. Disciplines cannot act as 'islands of knowledge in oceans of ignorance'[187] but have to find links among each other towards an interdisciplinary approach. Such an approach balances the ecology, economy, and social context of water. These are the three aspects that have to be addressed through the methodology. By involving stakeholders or representatives from these different aspects into the participatory process, the aspects can be incorporated into the understanding of or view on the problem.

The dynamics of policy problems require a flexible approach towards the process of information production, as stated above. This is necessary because the information needs of policymakers change over time, as the political attention for specific problems changes over time. These dynamics are depicted in the policy life-cycle. Information production therefore cannot set out a long-term program that is carved in stone for a prolonged period. Information production has to be continual to be able to change regularly as the policy issues change. On the other hand, however, the pursuit for sustainability requires a long-term perspective and the possibility to make future predictions. Long-lasting time-series of data are needed to enable that (see Box 1.3). The information production process therefore has to fulfill both these approaches to the extent possible; be flexible, but also continual. The role of the information producers in the methodology will largely be that of clarifying the problem and/or providing alternatives of scientific knowledge. They can also give support to find the balance between flexibility and rigor. The policy needs in this process in the end still need to come from the decision makers.

1.2.8 Some Definitions

A methodology for the assessment of information needs is described in this book, but what are information needs? The concept of information needs was studied in detail by Timmerman[78]. The resulting definition is:

> *"An information need is a precise question within a clearly defined context, specified to such an extent that an information producing system can be designed."*

Examples of information needs are provided in Chapter 5.

In this book, the terms information users and information producers are used as generic terms. There is no strict distinction between the two groups as, over time, everyone is a user and provider of information[31]. Nevertheless, people can be identified that in general produce information as well as people that in general have a need for information.

An *information user* all together is anyone who uses information. Information users are people from a broad spectrum of groups ranging from water managers, decision and policy makers to stakeholders like NGO's and the 'public at large' ("stakeholders are the parties who affect and are affected by a policy or program"[130]). To limit the wide variety in information users, in this book the central group for whom information is produced is the water management organization, more precisely the decision makers in such an organization. They are the ones who should specify what information they need and how they use this information. They are the information users that are targeted in this book.

Information producers are the people/organizations that collect, analyze and present information. In this book, the term 'information producers' refers to departments responsible for monitoring or information collection.

They report to decision makers in a water management organization. Information producers are the ones that manage the monitoring network and on the whole decide how the information is collected.

Box 1.22 Defining information

Dunn[130] used the following metaphor to describe the complexity of distinctions between data, information, knowledge, and wisdom: "In baking bread, the data used are molecules of carbon, hydrogen, and oxygen. In turn, the information at our disposal is starch, flour, water, and yeast. We have knowledge, however, only when we get to the next level, which is using the information to bake bread. Wisdom may be viewed as using knowledge to create a delicious croissant".

A vast body of literature exists on the definition of information and the differences between data, information and knowledge (Box 1.22)[125, 144, 188-192]. The concepts are not static; data can become information and information can become knowledge. Information is in general described as the sum of experience, understanding and available information. In knowledge management literature, distinction is made between explicit and tacit knowledge. Explicit knowledge is knowledge that is more or less easily captured, codified and communicated. It is transmittable in a formal language and can be stored in databases, libraries, etc. Tacit knowledge on the other hand is linked to personal perspectives, intuition, emotions, beliefs, know-how, experiences and values that are not easily articulated and shared[193-195].

Data, at the other end of the scale, is usually considered factual. For instance, an output from a device in a numerical form is data. The origin of data, how data is produced, is clear and if data is processed, the processing methods used are clear. It is possible to challenge the choice to collect specific data, the devices, and the methods and quality procedures by which data is produced. But once the quality of the data is considered acceptable, it is almost impossible to challenge the data itself. Vlachos, in studying transboundary water conflicts, concluded that in a transboundary situation data is rarely neutral; data collection itself is driven by values and is ultimately negotiable[196].

Information, finally, is often described as something that emerges from an interpretation or processing of data and that is exchanged or communicated. Different processing methods or different interpretations may lead to different information. The underlying data, however, do not change. In contrast to data, information is therefore to a certain extent subjective and can be challenged. The transfer of water information from scientists to decision makers is, in terms of the knowledge creation framework of Nonaka and Takeuchi[193] (see Box 1.23), a process where scientists externalize the knowledge they gain from the collected data and which they exchange and share with the decision makers.

The decision makers then combine and internalize this knowledge. In the terms of Nonaka and Takeuchi, information is best described as explicit (externalized or conceptual) knowledge[197]. This explicit knowledge is transferred when information is exchanged (Box 1.23).

Box 1.23 The process of creating new knowledge

The process of creating new knowledge is the movement between tacit and explicit knowledge over four modes of knowledge conversion. Knowledge socialization generates new tacit knowledge by sharing and exchanging know-how and past experiences, building a field of interaction. This yields what is called 'sympathized knowledge'. Knowledge externalization, triggered by dialogue, involves structuring tacit knowledge into 'conceptual knowledge' and making it available to other users. Knowledge combination generates new knowledge by combining and linking pre-existing explicit knowledge and bringing it together to produce new insight. This yields 'systemic knowledge'. Knowledge internalization finally converts explicit knowledge into internal, tacit knowledge: so-called 'operational knowledge'. Internalization happens when individuals, exposed to other people's knowledge, make it their own. People internalize knowledge by doing but also by looking at what other people have done in a similar context or by example[193, 195].

Box 1.24 The UNECE Aarhus Convention definition of environmental information

The UNECE Aarhus convention on environmental information[32] defines environmental information as any information in written, visual, aural, electronic, or any other material form on:

(a) The state of elements of the environment, such as air and atmosphere, water, soil, land, landscape and natural sites, biological diversity and its components, including genetically modified organisms, and the interaction among these elements;

(b) Factors, such as substances, energy, noise and radiation, and activities or measures, including administrative measures, environmental agreements, policies, legislation, plans and programs, affecting or likely to affect the elements of the environment within the scope of subparagraph (a) above, and cost-benefit and other economic analyses and assumptions used in environmental decision making;

(c) The state of human health and safety, conditions of human life, cultural sites and built structures, in as much as they are or may be affected by the state of the elements of the environment or, through these elements, by the factors, activities or measures referred to in subparagraph (b) above.

The term information will be used in this book as the general concept of something (knowledge, intelligence, or data) that adds to the knowledge of the receiver and is the tangible part of knowledge that, in the end, can make a change (see Section 1.2.9). Both the sender and the receiver add their ideas to the information, making information subjective (see Box 1.25). The term information in this study covers the range of types of information that is included in the definition from the Aarhus Convention[32] (Box 1.24).

Box 1.25 Communication between sender and receiver

Discussions in literature about communication often involve a model of a sender sending information through a channel to a receiver[198]. The model stems from a paper by Claude Shannon[199] who describes a model for electronic communication where a message coming from an information source into a transmitter. The transmitter sends out a signal that is received by a receiver that brings the message to its destination. The signal is influenced by a noise source (for a discussion on computer translation of data see Kirschenbaum[191]). This model is later applied for communication between people. All three elements and the transition from one element to the other can in this model cause communication to fail. Communication implies externalization of tacit knowledge into explicit knowledge by the sender. The mindframe of the sender determines how this takes place and a translation error can occur. Then the explicit knowledge is transferred through the channel, for instance when the sender writes a report. Depending on for example the writing abilities of the sender another error can occur. The channel itself may fail to deliver the full message, for instance when the receiver does not read the full report. The receiver finally has to combine the explicit knowledge and preferably internalize it, which may involve misinterpretation. On the other hand there may be an interpretation by the receiver that deviates from the intention of the sender. Exchange between the receiver and sender is therefore considered essential to ensure that the information is received or understood in the way it was intended. Nauen [200] noted that the impact of European research on society is largely determined by trust in the scientists concerned, the perceived relevance of the research thrust, and the communication capability. The impact of the communication is consequently linked to the source of the information, the mindframe of the receiver, and the supposed quality of the information.

1.2.9 The Use of Information in Water Management

Where Boogerd and others[107] studied the use of knowledge in decision-making this book takes the perspective of information in decision-making. A general appreciation in literature is that different people perceive information differently, which can lead to misunderstanding and difficulties in the communication between different actors[151]. People may perceive information in six different ways[201, 202]. The first perception, information as subjective knowledge (Box 1.26), describes information as directly related to the information user (equivocity) and is therefore part of all other perceptions.

Box 1.26 Information as a subjective concept

Capurro[188, 203, 204] describes information as a subjective concept depending on the interpretation of the receiver. He argues that information may mean the same (univocity), may mean something similar (analogy), or may mean something different (equivocity) for each person. The consequences of each of these suppositions will however take us too far from the purpose of this study.

The other perceptions that people may have of information are[205] (see Fig. 1.9):

- Information as useful data or as a 'thing' that is processed for a specific purpose and that is presented in a form that is meaningful to the recipient.
- Information as a resource that is attainable and useable and which accordingly can be managed similar to other factors of production.
- Information as a commodity with a specific value that is subject to trade.
- Information as perception of pattern that has a past and a future, is affected by various factors, and has effects.
- Information as a constitutive force in society that in itself is an actor affecting other elements in the environment.

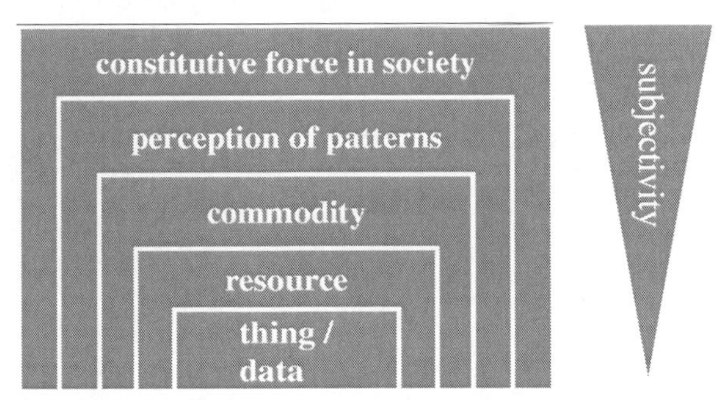

Figure 1.9 Different perceptions of information[205]

Box 1.27 The subjectivity of information

The subjectivity of information becomes apparent in the fact that what is just data to one person may be information to another person. For example a water level of 15.74 m above mean sea level on November 4, 1998 in the river Rhine at Lobith measuring station is merely data for most people. Specialists on this subject on the other hand know that this water level was among the top 10 highest levels in the 20th century. This illustrates that to the specialist, the data is supplemented with context derived from his/her knowledge, giving it meaning, relevance and purpose, and thus transforming it into information[104, 192, 206].

These perceptions are part of a hierarchy in which information as useful data is the most basic while information as a constitutive force is most encompassing[205, 207] (Fig. 1.9). The subjectivity of information (Box 1.27) increases in the range of characterizations from useful data to constitutive force. Information to the information user must at least be something he/she can use to improve his/her knowledge (combination) thus conforming to information as perception of patterns. If the information user is willing to change his/her perception and/or projected measures on the basis of information it can become a constitutive force (Box 1.28).

Box 1.28 Different perceptions of information

Haas[158] distinguishes between two perceptions of information. He argues that control over knowledge and information is an important dimension of power while the diffusion of new ideas and information can lead to new patterns of behavior. The first perception of information as a source of power is comparable to information to direct decision making, the second, information as a source of new ideas, is comparable to information to support decision making.

This becomes more tangible by looking at various (political) roles information can play in decision making for water management. Several roles are distinguished that are partly related to the perception of the water management problem, and partly related to the stakes involved. Some roles and uses of information in decision making are[205, 208]:

- Information as a legal obligation: In laws and regulations related to environmental management, usually monitoring obligations are included. This information has to be collected and reported. Information is in this sense a thing that has to be produced, regardless of any use that is made of it.
- Information as hideout or safeguard: In this situation, many activities are directed towards monitoring. It gives a sense of doing much useful work without having to take decisions. This approach is only useful in a situation where very little information is available and where information is a resource that has to have a certain volume before anything can be done with it (a longer time-series of data is needed for, for instance, statistical reasons). In other situations it becomes a way of postponing policy action, as described below.
- Information as a trading good: This is clearly the situation in which information is regarded a commodity that can be traded for something else. This viewpoint can be the start of cooperation when in a transboundary context the riparian countries agree to exchange information and in such a way can be the beginning

of building trust between the countries. Other situations, where information is treated as sensitive for security reasons or otherwise treated in a confidential way for commercial reasons or reasons of power, it may be an important reason for maintaining or increasing mistrust.

- Information to postpone decisions: In a situation where information is available but may give rise to discussion, stating that more information is needed will give the opportunity to postpone decisions, as stated above. Information is in this situation perceived as having a pattern, but the uncertainty is grasped to diminish the power of the information.
- Information to direct decision-making: The available information is in this situation split into useful and not useful information. Information that supports the desired outcomes will be put to the fore, while information that counters the desired outcomes will be discarded. There is a perception of pattern in this use of information, but the power of the information is locked up.
- Information to support decision making: This is more or less the ideal situation from the viewpoint of monitoring, in which the available information is the basis for the decision taken; it guides and supports the decisions. The information cycle builds on this role of information. In this notion, information can become a constitutive force.

Box 1.29 Different roles of information in the stages of water management

Grossmann[189] describes different roles of information in the stages of the cyclical process of transboundary water management that Mostert[209] describes. This framework will not further be used in this book but will shortly be described here as an example of how within a specific situation the role of information can change over time.

The first and often most difficult stage in transboundary water management is bringing the major stakeholders around the table. Exchange of information takes place at this stage for confidence development.

The second stage in transboundary water management is the negotiations. The relevant 'facts' have to be established, such as the natural river discharge, present use and projected demand, and several options need to be developed and assessed. Information is used here as a negotiation basis.

The third stage is the conclusion of an agreement. There is no information input here.

Implementation or compliance as the fourth stage is usually the responsibility of lower level governments and water users who have not been involved in international negotiations. Information is used here for verification.

When implemented, agreements result in certain environmental, social, economic, political and even cultural changes. Information now serves monitoring purposes to record these changes. The changes may be foreseen or not, but in any case they change the context of water management and may result in a new potential for development or conflict, in new negotiations and in new agreements. Information in the development potential is exchanged for instance about planned projects (notification) or for identification of issues for cooperation. One can imagine that depending on the level of cooperation between the riparian countries, information can be used from very different perceptions in this process.

For example the tendency of joint water management bodies in a transboundary situation to focus on joint monitoring efforts as described by Enderlein[210] may well be ascribed to the use of monitoring and the resulting information as either a hideout or to postpone decisions[211]. This data is then used as a trading good. Through assigning these roles to information, a joint monitoring program is portrayed by the riparian countries as being active in cooperating without actually having to take decisions. It is nevertheless often a first step in cooperation that over time leads to improved cooperation[35, 103, 105], where the information plays a supportive role in decision making[212]. Such a process requires building of mutual trust, which is a time-consuming affair[213].The perception of information is thus influenced by the interests of the information user as well as his/her role in the decision-making process.

Box 1.30 The rationality of decision making

Another approach that focuses on the question why people cooperate, and as such a process that interferes with the rationality of decision making, comes from comparative political science literature where the controversy circles around rational choices theories and cultural approaches. Rational choice explanations focus on individual cost-benefit calculations while cultural approaches stress history, habit and socialization. Rational actor approaches consider trust to be a product of effective institutional arrangements, while cultural approaches reverse this causal link, arguing that interpersonal trust is a precondition for democracy[106]. Although relevant for the decision-making process, this takes us too far from the purpose of this book.

A different perception of information thus leads to different use of information (Box 1.29). Decision-makers in general are inclined to consider information as either beneficial or dangerous and use the information they receive in accordance to their estimate (Box 1.30). Scientists generally consider information as something that can bring new insights and will value information as a support in confirming or rejecting their hypotheses. Such different perceptions are linked to different characteristics of the science process and the policy process as shown in Table 1.1.

Box 1.31 Different perceptions of information

Davenport[42] describes four models of information policies at different positions in a continuum between central and local control, which are linked to interests:

- Monarchy, where one individual or function controls most of the organization's information. This model compares to the situation in the monitoring practice of several EU directives that include reporting obligations stating exactly what parameters should be measured at what frequency at what location;
- Federalism: few information elements are defined and managed centrally; most is managed at local units. This model accounts for local particularities while setting some minimum requisites to information. The EU Water Framework Directive moves towards the federal model;
- Feudalism: business units control their own information environment. This model exists for instance in many transboundary water management situations, where a joint body is absent or hardly functioning. The riparian countries in this situation provide for their own information needs, while comparison of the water quality situation is not possible;
- Anarchy: individuals and small entities tailor information to their own needs. This model is present in many research projects, where data collection takes place without these data being made available to a wider audience than those directly involved in the research.

Table 1.1 Characteristics of science and government[111, 222]

Science	*Government*
Problem complexity reduced along disciplinary boundaries	Dealing with complex real problems
Based on experimentation	Not open for experiments
Looking for falsification	Looking for verification
Time needed for quality control	Quick results
Probability accepted	Certainty desired
Inequality is a fact	Equality desired
Anticipatory	Time ends at next election
Flexibility	Rigidity
Problem oriented	Service oriented
Discovery oriented	Mission oriented
Failure and risk accepted	Failure and risk intolerable
Innovation prized	Innovation suspect
Replication essential for belief	Beliefs are situational
Clientele diffuse, diverse, or not present	Clientele specific, immediate, and insistent

Information is considered an essential basis for decision making that feeds into a rational decision-making process and, although information is to some extent objective, the use of information is rather subjective as discussed in the previous section. The way information is interpreted and used is largely determined by people's norms and values. Norms and values are related to how we view the world, which is an assembly of our cultural background, professional training, character, experience, expertise, professional role, and so on[150, 214-221].

The three different explanations for the mismatch between scientists and decision makers that Boogerd and others [107] used to explain for the policy-science gap, namely the two communities debate, the mismatch model and the model of knowledge being politically and socially structured, are complementary to some extent and can largely be attributed to different mindframes. Improved communication and interaction between the groups is in literature generally regarded as the preferred approach to deal with the mismatches[12, 37, 223-226]. This book therefore builds on the idea that to improve communication between scientists and decision makers, both parties should be aware of the fact that their opinion is limited by their mindframe and they should be willing to try to get an understanding of the other party's mindframe[227]. This implies that people negotiate with each other over their values, their goals, their differing interests, etc. and develop a common view on the issue at hand. Jim Woodhill[228] phrases the situation as follows: "Water resources management requires negotiating conflicts and differences between different stakeholders. Conflict is often a clash of paradigms. People act and rationalize things in a way that does not make sense to others because they are operating with a different set of assumptions, values and beliefs. In resolving such conflicts, those involved need to make their paradigms explicit and see others' paradigms—for which facilitation is often critical". Developing a common view can only be achieved by interaction and close communication between the actors. And this is basically what is promoted within the concept of social learning literature[162, 217, 228-230]. The methodology as described in this book also builds on this concept of social learning.

1.3 POLICY PROBLEMS AND APPROACHES TO SOLVE THEM

A range of approaches to target policy problems are described in literature. This section describes some of these approaches, how policy problems can be structured in preparation to solve the problems, and the approaches that different scholars have described to solve policy problems. It is important to be knowledgeable about these approaches, as they have been applied in developing the methodology for the purpose of translating policy problems into information needs and incorporated into it. And for proper understanding of the methodology, some understanding of the underlying ideas is required.

1.3.1 Approaches towards Problem Solving

Formulating the policy problem and building a common view or understanding of it is the first and probably most important step in problem solving in general. "The problem definition ramifies throughout the problem-solving process, reflecting values and assumptions, determining strategies, and profoundly impacting upon the quality of solutions"[231]. The importance of adequate problem exploration is illustrated by a study performed by Interaction Associates in 1986 which claims that 90% of problem solving is not spent on the actual problem but on[231]:

- Solving the wrong problem;
- Stating the problem so it cannot be solved;
- Solving a solution;
- Stating problems too generally;
- Trying to get agreement on the solution before there is agreement on the problem.

For this reason, solving the problem starts with defining the problem. If this is properly done, it enables avoiding errors of the third kind[172]. Or as Perry and Vanderklein state: "The key to effective management is learning to ask the right questions"[20]. Hisschemöller and Hoppe state that problem defining and problem solving are not separate stages in the policy process and that these policy strategies have their implications in policy-analysis[169]. They differentiate between four types of policy strategies for problem solving. Each of them is related to one of the four types of policy problems they distinguish:

1. The first strategy is the rule strategy. In this strategy, authorities impose measures to solve the problem. This strategy is suitable in a structured problem situation when the ends (norms and values) and the means (relevant knowledge) of the problem are clear in the viewpoint of the authority.
2. In the negotiation strategy, multiple actors are involved; organized social groups and government representatives articulate their positions on the issue and try to reach consensus. This strategy holds best in moderately structured problems when the relevant norms and values are clear and negotiation is about the relevant knowledge.
3. The third strategy is the accommodation strategy where a small group of experts is involved in problem solving. The dispute concentrates on the discordant norms and values in an ill-structured 'means' problem situation.

4. The preferred policy strategy in an unstructured problem situation is the learning strategy. This strategy involves multiple actors and has a high level of public participation. There is an almost complete equality among the actors based on the idea that "an expert is not a special kind of person, but each person is a special kind of expert"[169]. Especially this last strategy concentrates on problem structuring.

The level of problem structuring consequently depends on the type of problem at hand. Not only is this the case in policy-analysis but also when information has to be provided as Gregory states[232]: "Once the right problem has been identified, the basis for future consultation is the structure provided by specifying the relevant objectives, agreeing on how they will be measured, and creating an initial set of alternatives for consideration". This approach of problem solving is translated into information production when the methodology as described in this book was developed. The step of specification of information needs from the information cycle (see Chapter 2) is in this approach largely the step of formulating and structuring the 'information' problem; what is the information needed?

The learning strategy and negotiation strategy as described by Hisschemöller and Hoppe, will be the most important strategies for specification of information needs, as the methodology largely deals with clarifying the relevant knowledge. A structured approach towards the problem is a promising route to avoid errors of the third kind[172]. Implementation of such a structured approach is a challenging effort, in part because it is different from what decision makers or stakeholders have come to expect[232]. But how can problems be structured?

1.3.2 Structuring of Problems

How a problem is defined determines the understanding and approach to the problem and thus the possible solutions to the problem [231]. Defining a problem starts with structuring or framing the problem, as described in the previous section. Proper exploration of the different aspects there are to the problem can improve the quality of solutions. Experts in the relevant field usually need little effort to conceptualize the problem and its implications. They may however be biased by their mindframe, which can make them overlook unique aspects of the problem. Novices in the relevant field, on the contrary, will have difficulty to oversee the problem. For this reason, it is important to structure the problem in a way that explores the problem without skipping peculiarities of it. Bardwell proposes two elements to frame the problem[231]:

1. Managing the process. This element relates to the organization of the process of framing the problem.

2. Organizing the problem: building structure. This element relates to the cognitive map (a cognitive map is a representation of the relationships, which are perceived to exist among the elements of a given environment[233]), structure or mental model that is used for the problem.

To manage the process of problem framing (or problem structuring) one has to have a model of the process at hand. Especially in the case of ill-structured or unstructured problems a model offers a means to "approach problems that might otherwise have been avoided, forsaken, or just poorly solved"[231]. The problem-solving effort involves, according to Bardwell, the following stages[231]:

1. Building an understanding of the problem: defining the problem-space;
2. Establishing initial criteria for the goal;
3. Searching for solutions;
4. Deciding among solutions;
5. Evaluating progress: comparing initial goals to and monitoring of the solution.

These stages form a model of the process. The final stage feeds back to the second stage where the solutions are compared to the stated goals. The last four stages often dominate the problem-solving effort. Nevertheless, the first stage, understanding of the problem, is the essential one. Unfortunately this is the one that is often left out. Keeney[234] refers to this tendency as alternative-focused thinking as opposed to value-focused thinking (see Box 1.6). The searching for alternatives—or solutions in the terms of Bardwell[231]—usually gets the focus in problem solving while the actual problem is seldom really clear yet. The understanding of the actual problem is, as stated, more important because: "the problem definition implicitly embodies preconceptions and assumptions (mindframes) that underpin how one approaches the problem". These mindframes are related to the values of the people involved. People's values are usually implicit, while defining the problem requires to make them explicit. But the problem definition steers the strategies and actions taken to address the problem. It therefore influences all following actions and decisions. Organizing the problem is consequently the other element of problem framing that compels ample attention. To structure the problem framing, a model of the problem at hand is required next to a model of the process.

Dunn[130] describes a refining process that can be performed through application of different strategies. He provides several policy models as simplified representations of selected aspects of a problem situation. These models help to simplify problems, this at the same time being their major weakness. The two main forms of models are descriptive and normative models. Dunn thereby focuses on mathematical models. But looking at models in general, like statistical or conceptual models, this distinction is also useful. Descriptive models explain the causes and/or predict the consequences of policy choices whereas normative models also provide rules and recommendations for optimizing the realization of some value.

As Bemelmans[104] noted, people tend to refer to existing situations and are not able to come loose from it. Bardwell[231] provides different strategies to deal with this, among others, through limiting the information by simplifying the problem (without making it too simple), generating imagery through establishing patterns and a perspective for looking at a problem, and developing metacognition; monitoring, watching, and guiding one's own problem-solving process. De Bono[235] promotes the concept of lateral thinking as a means to open up new insights. He subscribes to the notion of Bemelmans by stating that "The main purpose of mind is to be brilliantly uncreative". To overcome this narrow-mindedness, de Bono introduces lateral thinking, which is largely a way of taking a different angle of view towards problems. He promotes the use of humor and provocation as means to invoke lateral thinking. A humorist often pushes the audience in a certain direction but with the punch line shows a totally different view of the situation through a double meaning of a word or a totally different situation than expected. Provocation does a similar thing by taking a standpoint contrary to the normal line of reasoning.

Keeney[234], to finish with, states that improved understanding of the decision context can be achieved by structuring the objectives. A clear distinction between fundamental and means objectives can be made by linking the objectives and creating a hierarchy of objectives. Through such a hierarchy of objectives, omissions in the objectives network become clear. Construction of an objectives hierarchy is a process of identifying the overall (highest-level) objectives and clustering of lower-level objectives until a sufficient structure is created.

1.3.3 Solving Policy Problems

In general, a problem is a discrepancy between the actual situation and a desired situation. As policy problems are often very complex, the actual situation can be described in an objective manner only to a certain extend. The desired situation is on the other hand largely dependent on values.

In general there are three levels of attitude that steer human behaviors: beliefs, values, and opinions[236]. Beliefs are the strongest feelings that are connected to moral reasoning. Values are based on these beliefs. People's values to a large extend steer their actions. The way people, for instance value the environment determines how they compose our lifestyle. This value of the environment in turn is based on our beliefs with regard to human superiority and ecological integrity. Opinions, finally, are based on our values. This is the way people think they should behave. If people are convinced that they can be more environmental friendly without losing comfort, for instance by separating glass from other garbage, they will change their opinion without changing the underlying value. People can also change their values, like for instance the book "Silent spring" did for many people in the 1970's. People's beliefs however, are very hard to change.

As values differ and are usually implicit, it is not easy to come to a good definition of a problem. Then, as discussed in the previous section, if the problem is not clear, it will be difficult to provide information to support solving of the problem. Policy problems are dealing in particular with balancing values. Problem solving, as a result, often focuses on solving the wrong problem. In dealing with policy problems it is therefore necessary to bring structure. The essence in problem structuring is that there is an obvious need to define and structure a policy problem before any action is taken. The classification of Hisschemöller and Hoppe of 'unstructured', 'ill-structured' and 'moderately structured' problem situations can help to identify the ways in which structuring is needed.

Several approaches to problem structuring are described in literature. Bardwell[231] emphasizes the importance of a model of the problem situation but indicates that the process to structure the problem is as important and discusses several strategies for defining the problem:

- Staving off solutions: People have a tendency to look for solutions and may pass over the problem definition too quickly. By emphasizing on exploring the interests—the needs and concerns —shared or at least non-competing interests may be found. After this, the problem may be changed in its definition.
- Limiting information: People are limited in understanding and absorbing information and more information may turn the attention from what is important to what is unimportant. Therefore, the parties may agree on what information will be used, thus limiting the amount of information.
- Choosing levels: The scale or level at which the problem is approached is chosen. His may avoid problems becoming too big and overwhelming or too small and negligible. Some factors to consider are:
    - Fit: the skills and abilities of the participants in the process must be adequate for the task.
    - Linkage: The linkages among levels provide a context in which actions are part of a larger framework.
    - Personalization: Issues that relate to one's own circumstances tend to offer more tangible results.
- Generating imagery: People should be provided with possible approaches that will help in coping with the problem.

- Developing metacognition: Each of these strategies implies simplification or decreasing of the problem into manageable parts. Some sort of 'meta' structure should be included to help fit the parts together.

Dunn[130] describes the process of problem structuring and adds several types of models and methods to structure problems as described in Section 1.3.2. Gregory[232] lists several elements that can be used as viewpoints, starting from which it is possible to structure the problem and as such provides a sort of model of a generic problem situation. "Once the right problem has been identified, the basis for future consultation is the structure provided by specifying the relevant objectives, agreeing on how they will be measured, and creating an initial set of alternatives for consideration". He provides the following key elements for structuring and delineating problems to come to smart choices:

- Problem: Define your decision problem to make sure to solve the right problem.
- Objectives: Clarify what you are really trying to achieve with your decision.
- Alternatives: Create better alternatives to choose from.
- Consequences: Describe how well each alternative meets your objectives.
- Tradeoffs: Balance objectives when they cannot all be achieved at once.
- Uncertainty: Identify and quantify the major uncertainties affecting your decision.
- Risk tolerance: Account for your appetite for risk.
- Linked decisions: Plan ahead by coordinating current and future decisions.

In this list of elements, the first element can be seen as an overall delineation of the problem. The other elements are different view angles to approach and structure the problem. Some elements are interlinked but they all feed back to the first, the problem at hand. Here again, partitioning of the problem is done, through limiting of the information by looking at specific aspects of the problem. The overall problem itself provides the meta-structure, the others help to define and specify the problem.

De Bono[235] advocates a noncompliant view towards problems. He promotes the concept of lateral thinking as a means to open up new insights. To overcome narrow-mindedness, de Bono introduces lateral thinking, which is largely a way of taking a different angle of view towards problems. He promotes the use of humor and provocation as means to invoke lateral thinking. A humorist often pushes the audience in a certain direction but with the punch line shows a totally different view of the situation through a double meaning of a word or a totally different situation than expected. Provocation does a similar thing by taking a standpoint contrary to the normal line of reasoning. Keeney[234], finally, links the structuring of problems to the policy objectives, thus explicitly linking back to the values that also Bardwell highlights as described in Section 1.3.2 (also see Box 1.6).

Box 1.32 The Logical Framework Analysis (LFA)

The Logical Framework Analysis (LFA) tries to create a logical sequence linked to realization of the goals. The first concept of LFA is Project Logic, the statement of how the main elements of a project will lead to the desired success. There are usually four parts of the Project Logic: Inputs, Activities, Outputs and Purpose. Each part will lead to the next part up the hierarchy: IF these Inputs are available, THEN these Activities can take place. Next, IF these Activities have taken place, THEN these Outputs will be produced. Finally, IF these Outputs are achieved, THEN this Purpose will be fulfilled. LFA appears to be assuming a direct linear inevitable sequence of events, but this is only partly the case. The next thing that is required is an explanation of why it is likely that the logical sequence will actually take place. More important perhaps, it asks what might cause the logical sequence to fail. This aspect in the logical framework is called 'Necessary conditions for success' or 'Important assumptions' . This is one of the most useful parts of LFA since it obliges the project planners to examine the assumptions that they are making[7, 123].

Important features of all these approaches are that the problem is delineated. Two elements are important here; namely organizing the problem and managing the process. The following step is then to limit the amount of information that people have to deal with. Dunn suggests classificational analysis and hierarchy analysis and LFA (Box 1.32) builds a hierarchy while Gregory proposes alternatives. The latter are unwanted according to Keeney, as they blur the sight on fundamental objectives and the values that lie behind them. No choice is made between these notions here but a closer look at the importance of values is provided in the next section.

Looking at the actors in the process of solving problems, it is found that in many instances multiple actors have to be involved in a learning or negotiation strategy. The role of information producers in this process is to—collectively—identify or clarify the problem and to advocate specific viewpoints. If the process runs into disputes over norms and values, the nature of the process will change into an accommodation strategy where the information producers act as mediators[168].

Values direct the problem formulation and, as a consequence, largely determine the information needs to support problem-analysis and problem solutions. As policy problems are at the center of policy analysis, values should be the driving force for our problem definition. One of the key elements in policy-analysis is formulating problems as part of a search for solutions. Values are in this respect the key to the solutions as they determine the final goal. Problem structuring as a consequence requires clarifying values of the stakeholders (Box 1.33).

Box 1.33 Value-focused thinking[234]

'Values are what we care about' . Value-focused thinking as described by Keeney essentially consists of two activities: first deciding what you want and then figuring out how to get it. The first step of value-focused thinking is to broaden the decision situation and define it more carefully. This should generally be the approach in developing an 'information system' to support water management, as displayed in the information cycle (see Chapter 2). Instead, problem analysis usually focuses on choosing among alternatives, first figuring out what alternatives are available and then picking the best of the lot. Alternatives however should be the means to achieve the more fundamental values and should only be the second step after deciding what is really wanted; what the objectives are. The major shortcomings of alternative-focused thinking are that (1) viable alternatives, possibly much better than the alternatives considered, are not identified; (2) the objectives identified are often only means to the consequences that are of fundamental concern; and (3) there is not a logical match between alternatives and objectives. In short, alternative-focused thinking is too narrow.

The general principle of thinking about values is to discover the reasoning for each objective and how it relates to other objectives. The achievement of objectives is the sole reason for being interested in any decision. And yet, unfortunately, objectives are not adequately articulated for many important decisions. The process of identifying objectives requires significant creativity and hard thinking about a decision situation. Thinking about values can be helpful in:

- Uncovering hidden objectives: Bringing values to consciousness allows uncovering hidden objectives, objectives one didn't realize one had.
- Guiding information collection: Once someone's values are specified, one should then collect information on alternatives only if the information will help judging the alternatives in terms of achieving those values.
- Improving communication: The language of value-focused thinking is the common language about the achievement of objectives in any particular decision context. It is not the technical language of many specialties. This basis in common language should facilitate communication and understanding.
- Facilitating involvement in multiple stakeholder decisions: Many decisions, including those categorized as bargaining or negotiation, involve multiple stakeholders who must interact to produce decisions. Value-focused thinking can contribute to the productivity of such interactions.
- Interconnecting decisions: Alternatives chosen in different situations should not work as cross-purposes.

- Evaluating alternatives: If the value model is not based on sound judgment and logic, the insights based on evaluation with that value model cannot be sound.
- Creating alternatives: It may be much more important to create alternatives than to evaluate readily available ones.
- Identifying decision opportunities: Often, instead of sitting and waiting, it may be preferable to identify decision opportunities, that is, opportunities to better achieve our over-all values by formulating a decision situation.
- Guiding strategic thinking: The strategic values can suggest when and where potentially productive decision opportunities may be lurking. They also suggest the objectives for those decision opportunities, which are more specific than strategic objectives.

All this however requires values to be made explicit, which in practice will be a laborious task off the well-known track. Nevertheless, given the above, it is worthwhile exploring the various issues connected with values.

An objective is a statement of something that one desires to achieve. An objective indicates the direction in which someone is striving. A goal differs from the objective in that a goal is either achieved or not[237]. Keeney distinguishes between two types of objectives; the fundamental objective that characterizes an essential reason for interest in the decision situation and a means objective that is of interest in the decision context because of its implications for the degree to which another (more fundamental) objective can be achieved. As an example, an objective in water management may be to achieve high transparency of the water. This is a means objective, because it is of interest only because of its implications for the quality of the ecosystem. The underlying fundamental objective is that a good water quality is needed for good ecosystem functioning. Transparency of the water increases as the ecosystem functions better.

Fundamental objectives are essential to guide all the effort in decision situations and in the evaluation of alternatives. The decision context and the fundamental objectives together provide the decision frame. Values of decision makers are made explicit with objectives. The fundamental objectives are the basis for any interest in the decision being considered. These objectives qualitatively state all that is of concern in the decision context. They also provide guidance for action and the foundation for any quantitative modeling or analyses that may follow this qualitative articulation of values. If objectives are missing or vague, alternatives may not be recognized, time and energy may be wasted collecting unnecessary information while useful information may be ignored, and communication about the pros and cons of the alternatives will suffer.

An objective is characterized by three features: a decision context, an object, and a direction of preference. To make an objective operational, these three features are needed. As an objective is measured in terms of the attributes[237], or as attributes are the criteria linked to objectives, the object in the decision context is described through the attributes. The objective should best be stated in terms like "Twenty percent reduction of water pollution in the next five years", or "No more interruptions in the intake of drinking water within two years". In this objective, there is a decision context ('overall water quality', 'drinking water'). The objects or attributes are pollution or interruption of intake. Then the direction ('percentage reduction' or 'no more' or 'less than') is included in the specification. Next to that an element of time ('.... within two years') should be included. "By the year 2000 the ecosystem of the river Rhine should be improved to such an extent that higher species, such as the salmon, may become indigenous"[238] is a statement that in a way can be measured. There is a decision context (ecosystem of the river Rhine), an object (salmon), an element of direction (a more or less stable population), and an element of time (the year 2000). In this case, the object should be expressed in terms of specific water quality attributes like oxygen, temperature and toxicity, and to physical factors like suitable spawning grounds and possibility of migration. When objectives are quantified, in the end the information need can be converted into an information strategy[239].

Problem definition is important to be able to structure policy problems. This is not easy because different people have different values and consequently come to different definitions of the problem. And people are inclined to think about solutions rather than problems. As each actor and party in a policy problem considers its own solution superior to the others in terms of problem solving, discussions may be focused too much on pros and cons of solutions and weighing these. If however people would focus better on their objectives, it may well be that solutions can be found that meet a majority of the objectives and as a result satisfy more parties then would be possible in the situation where solutions are traded. This is also largely the distinction that Turner[184] makes between optimizing one goal at the cost of others and satisfying all the different goals.

By probing the values of the different parties, the fundamental objectives may become clear and these in turn will enable a shared problem definition. Values and the way they are expressed are essentially based on mindframes. Especially in a transboundary water management situation expression of values and fundamental objectives is difficult because of the different cultures in different countries. It becomes even more important in such a situation to try to probe the values of the parties involved and use these as the basis for the problem definition. This approach is thus an important element of organizing the problem and consequently forms an important part of the methodology, as will be described in the next chapters.

1.4 EXERCISES

- Make an overview of the water monitoring network developments in your organization. This includes describing the developments in number of monitoring locations, the number of parameters, and the measuring frequency of monitoring networks over the years.
- Determine how the information collection is organized. This is done by answering questions like "who determines what information is collected?", "who is responsible for collecting the information?" etc.
- Who will be the main receivers of the information once it is produced? Identify the people and/or departments in your organization that use the information. Keep in mind that this may be a very diffuse group.
- Do you experience a 'water information gap'; is there a feeling among information users that they don't get the right information? If yes, determine what the main problem is; is there an issue of legitimacy (for example, the department that collects the information is insufficiently trusted), credibility (for example, do the methods used (for instance network composition, sample collection, analysis methods, calculation methods, etc.) live up to (inter)national standards?), or salience (are the information needs attuned with the information users and do they reflect the issues that concern the users?), or all three?
- Do you see ways of bridging the 'water information gap'? Based on the answers to the previous question, what would be the first thing to do to overcome the gap?
- Give an overview of the water management situation; what are the main policy issues that are in need of information? Where is the focus of the information collection?
- What phase of the policy life-cycle are these different issues in? Are these issues emerging or are they linked to well-known policies in the control-phase?
- How is the information transferred to the policymaking domain? Is it in the form of reports, fact sheets, presentation, workshops, or other means? And is there a check on if the information as conveyed is received as intended?
- How is the information as produced used in the policymaking process? Can you identify where the products as produced in the information production process end up in the policymaking process?
- Why would the learning strategy and negotiation strategy be the most important strategies for specification of information needs?

2

The Link between Monitoring and Water Management

This chapter deals with the water information production process. It discusses the information cycle that is developed as a framework to describe the process of producing water information. It also elaborates on the steps distinguished in that process. The cycle signifies that the process is an iterative process in which each iteration leads to improvements in the information produced. Such improvements are initially needed to better tune the information to the needs of the information users but, over time, is also necessary to account for the changes in policies, technological developments, and other external factors that compel to update the information producing system. After studying this chapter, the reader will have obtained the following abilities:

- Understand what is monitoring
- Understand the role of the information cycle as the link between water monitoring and water management
- Understand the steps in the information cycle
- Be able to apply the information cycle to the monitoring process in your own organization.

2.1 WHAT IS MONITORING?

This chapter describes the water information production process. Water monitoring as an important information production process is in much of the literature on water management described as a rather technical task. It is often presented as a one-way street from policies to information production where monitoring objectives should be related to the applicable policy and the resulting information should be relevant for the water manager. However, little attention is paid to ensuring that these conditions are met. For one thing, although the importance of specification of information needs is generally acknowledged, how this specification should take place is hardly elaborated. If practiced, it is done from a technical perspective. In such a situation, usually some remarks on the relevance of the information produced for policy making are included in monitoring network design. The actual effort usually goes into determining about the legal obligations and other, more technical, considerations that steer the network development.

This book aims to overcome the problem that information needs are rarely specified by offering a straightforward approach towards determining information needs that include the information users' perspective. The approach requires involvement of water management decision makers and water monitoring people. The book thus aims at building a two-way bridge between water management and water monitoring. The previous chapter described the water management process and the issues that play a role in developing policies. This chapter aims at creating a better understanding of the monitoring side of the bridge to be able to link up to that process.

2.1.1 Defining Monitoring

Monitoring is described in literature by a multitude of definitions, each giving its own purposes and limitations to monitoring, depending on the goals to be achieved. One definition is given here, from the Guidelines on Monitoring and Assessment of Transboundary Rivers[179]:

> *"Monitoring is the process of repetitive observing, for defined purposes, of one or more elements of the environment according to pre-arranged schedules in space and time and using comparable methodologies for environmental sensing and data collection. It provides information concerning the present state and past trends in environmental behavior".*

This definition is immediately followed by the definition of assessment[179]:

> *"Evaluation of the hydrological, morphological, physico-chemical, chemical, biological and/or micro-biological state in relation to reference and/or background conditions, human effects, and/or the actual or intended uses, which may adversely affect human health or the environment".*

Information in these definitions of monitoring focuses on water quality and quantity. Environmental information, however, was defined in a much broader sense in Chapter 1 and incorporates among others societal and economic elements as well as policies and administrative measures and this book will build on this broader definition. In line with the broad definition of information from the UNECE Aarhus Convention[32] (Box 1.24), Dunn[130] describes five types of policy-relevant information:

1. Policy problems: What is the nature of the policy problem?
2. Policy futures: What policy alternatives are available to address the problem and what are their likely future outcomes?
3. Policy actions: What alternatives should be acted on to solve the problem?
4. Policy outcomes: What present and past policies have been established to address the problem and what are their outcomes? and
5. Policy performance: How valuable are these outcomes in solving the problem?

Dunn[130] defines monitoring as providing information on what he calls the policy outcomes. Contrary to the abovementioned definition of monitoring that includes all elements of environmental information, Dunn with this definition excludes monitoring from other policy-relevant information. Despite this limitation it is important to mention here that monitoring performs at least four major functions in policy analysis according to Dunn:

1. Compliance: do the actions taken result in compliance with standards?
2. Auditing: are the measures actually taken?
3. Accounting: do the measures result in improvement?
4. Explanation: how are the processes in the field?

These four functions must be present to enable evaluation of the effectiveness and efficiency of the policies[130]. Environmental monitoring mostly fills in two of the functions; compliance monitoring (for instance, testing against standards) and accounting for improvements, usually in the form of trend monitoring. Monitoring for explanation is often not among the objectives of monitoring networks, but does have its influence on research aiming at understanding the processes in the field. The auditing function is often only done on an ad hoc basis. In 2003, however, the Dutch organizations responsible for water management (on national, provincial, and communal level as well as the water boards)[240] jointly agreed, among others, to collect information on implementation of measures on a regular basis.

For the purpose of this book the abovementioned definitions of monitoring and assessment will be extended here to include the full range of environmental information as defined in the UNECE Aarhus Convention[32], with the limitation of being relevant for water management. Monitoring should therewith fulfill the four functions of compliance monitoring, auditing, accounting and explanation as described above.

Monitoring cannot be separated from assessment; without assessment, monitoring only produces data, not information. Therefore, when monitoring is mentioned in this book, it will be regarded as including the assessment of the collected data. As is depicted in the information cycle (Section 2.2), the information production starts off from the water management process and ends in the water management process. This implies that the information as produced is directly usable in that process and therefore includes assessment of the data.

On the other hand, not all information gathering is monitoring[241, 242] (Box 2.1). Research and modeling, for instance, are activities that are closely inter-linked with monitoring and all three are required to detect and manage environmental change. Information collected for research however seldom reaches the water management domain but usually stays in the science domain. The explanation function of monitoring is therefore usually limited. Modeling on the other hand needs monitoring information to validate the models. Nevertheless, with the currently available information, modeling is now more and more used to fill in part of the monitoring efforts. The essential function of modeling is to explore possible futures. As such it can be an important element of the information production process.

Box 2.1 Monitoring as 'informal walking in the rain'

A somewhat deviating notion towards monitoring is described by MacDonald and Smart[243]. They state that monitoring should be considered as a continuum that ranges from 'informal walking in the rain' to carefully replicated, quantitative studies. Depending on the objectives and context of the monitoring project the approach is more or less subjective, very formal, or anything in-between. The formal approach is necessary to provide more detailed and defensible results. The informal approach can be helpful to obtain a first impression upon which more formal monitoring projects can be built. This approach can also be found in the stepwise testing strategy and phased approach as described in the UNECE Guidelines on Monitoring and Assessment of Transboundary Rivers[179]. The adaptive approach towards monitoring is an important concept when looking for more flexible monitoring, for instance in case of monitoring for topical situations like flood situations or emergent hazardous substances[244]. Monitoring in this book is not limited to the formal, well-organized and repetitive data collection, but also encompasses more informal, qualitative types of information gathering.

2.1.2 Design of Monitoring Networks

How is the monitoring design usually described in literature? Several models for the design of monitoring networks are found[13, 128, 245]. One clear statement in most of this literature is that the design of monitoring networks is done on the basis of general monitoring objectives that are directly related to water and its use. General monitoring objectives are for example[90, 126, 246-250]:

- Identification of baseline conditions in the watercourse system.
- Detection of any signs of deterioration in water quality.
- Identification of any water bodies in the watercourse system that do not meet the desired water quality standards.
- Identification of any contaminated areas.
- Determination of the extent and effects of specific waste discharges.
- Estimation of the pollution load carried by a watercourse system or sub-system.
- Development of water quality guidelines and/or standards for specific water uses.
- Development of regulations covering the quantity and quality of waste discharges.
- Development of a water pollution control program.
- Quality control of water used for a specific purpose, for instance drinking water.
- Description of water quality, including trends in time and space.
- Evaluation of the effectiveness of a water quality management intervention.
- Identification of potential benefits from proposed or alternative remediation options on which decisions for investment options can be based.

Derived from these objectives the network is designed, specifying what (determinand) has to be measured where (location) and how frequent.

All objectives given above have a technical/scientific focus on status and trends. Note that the auditing function is not addressed in either of these objectives. Furthermore, only the latter two bullets implicitly include looking at measures and their effectiveness (accounting function). This latter information is nevertheless usually very relevant for policymakers, as will be discussed in the following chapters. The relationship between these monitoring objectives and the water policy problem situation is weak. Nevertheless, it is generally acknowledged that one of the important reasons of the insufficiency of (monitoring) information is that interaction between information users and information producers is poor[104, 107, 111, 115, 123, 251] (also see Chapter 3). To overcome these shortcomings, information users should be included in the development of an information producing process[252].

Ward and others[92] take an initial step towards involving water management and the decision makers. They define the first step in the design of a monitoring network as the evaluation of information expectations. This entails looking at water quality goals, water quality problems, management goals and strategy, the monitoring role in management, and monitoring goals. The monitoring system linked to this principle is defined as containing a number of tasks to convert a sample of water in the environment into accurate understanding of it[128]. The starting point is the water quality situation in the environment. The distinguished tasks are to collect samples, analyze these samples in the laboratory, and store, analyze, and report the data. When this resulting information is utilized, an accurate understanding of the water quality conditions is obtained. The various monitoring frameworks in literature (also see Box 2.2) are suitable as design tools for monitoring networks. Often in these models, the input of the different actors is described. The frameworks are however rather research oriented and rarely explicitly refer to policy objectives, which is a major weakness when linking information to policies.

Box 2.2 Monitoring frameworks in literature

A range of monitoring frameworks can be found in literature, among others, Bartram and Helmer[119], Bosch and others[253], Buishand and Hooghart[254], Mancy and Allen[90], Meybeck and others[120], Mulder and others[255], Passino[256], and USGS[126].

Cofino and others[257-259] developed the quality cycle (or quality spiral) that turns the linear framework into a loop that improves with every turn of the loop. They thus turn attention to the explicit and iterative process necessary for the efficient design and execution of monitoring projects. MacDonald[124] builds on this cycle and states that managers in consultation with technical staff should identify the information objectives. The managers are the ones to specify what information is needed. The technical staff (information producers) should be involved in this as they are the ones to translate the information questions into monitoring networks. A closer link is made to the policy objectives in this way.

2.2 THE INFORMATION CYCLE

The information cycle (Fig. 2.1) is used in this book to explain the process of information production. The information cycle is based on the quality circle of Cofino[257] (Fig. 2.4) and includes elements of the model of Ward and others[92] and the monitoring feedback loop of MacDonald[124]. It also targets the weaknesses in these models by making an explicit link to the water management process and the policy objectives included in it.

The first step in the quality cycle (and in the model of Ward and others) is evaluation of the information expectations, in which Management as displayed in the monitoring feedback loop of MacDonald[124] plays an important role. Management is however not necessarily involved in developing the monitoring plan, as this is a more technical task. On the other hand, an information need does not unavoidably lead to monitoring. Then, as Ward[260] emphasized, mere reporting of the information produced is not enough to satisfy the user, there should be ample attention for and evaluation of the use of the information. These considerations combined have led to the information cycle.

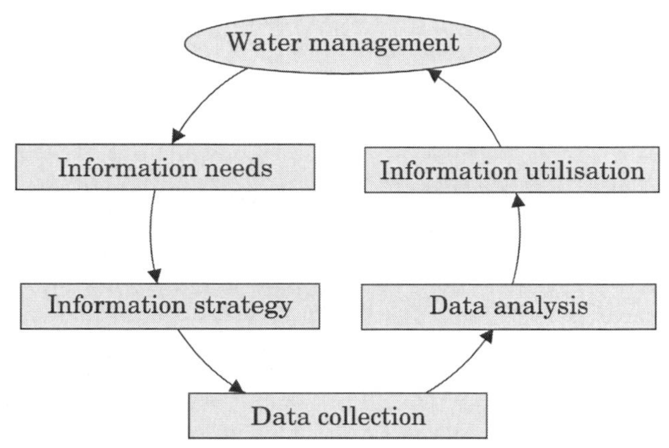

Figure 2.1 The information cycle[263].

The information cycle describes the essential elements in the process of information collection, which is inextricably bound up with the process of water management, including the policy process. The nature of the information needs will vary over time[261] requiring the process to be repeated. The resulting information is the sum of a chain of activities where the strength of the cycle is determined by its weakest link[262]. The cycle is easy to understand, describes the distinct production elements and mutual connections in which the position of the actors is clear, and has an intrinsic feedback loop.

In the information cycle, going from information required to information obtained, the following elements are distinguished[58, 264]:

1. Information users, as part of the information cycle, should, in co-operation with information producers, decide upon the characteristics of the information that is needed: the information needs. This element represents the link between water management and the information production process and is the main subject of this book;
2. Information producers will, in co-operation with information users, decide upon the best way (i.e. strategy) to collect information with a specified (required) quality in the most efficient and cost-effective way;
3. A monitoring plan can be the outcome of the information strategy, but this is not inevitable as other sources of information

exist; other programs to collect data can be developed in this step. The information cycle can therefore be used as a generic framework for designing information collecting systems;

4. The actual collection of data is the next step in the information cycle. Depending on the type of data and information needed, the data may be collected through, for instance, monitoring, models, or literature survey;
5. The collected data are analyzed and the results are interpreted relative to the information needs. Information statements are made on this basis. The goal of the data-analysis is to mold the information in a form that the information users can utilize;
6. The resulting information is presented and transferred to the information users in a proactive manner. Science is linked again to water management in the activity of information utilization. This again, requires a dialogue to ascertain that the information is interpreted in a way that reflects the actual situation.

The repetitive character of monitoring over time is reflected in the cyclic character of the model, where every now and then the contents of the information to be collected are reconsidered and improved where necessary.

Starting an information producing process is not something that is done from scratch. There is always an existing form of information production that will influence the process at hand. The information cycle as it is portrayed with activities linked by arrows gives the logical flow and sequence of information and activities. As such it is a simplification of reality. Nevertheless it is important to realize that in applying the information cycle, at each stage, consultation with users is essential to make sure that the information will satisfy or exceed expectations[265]. On the other hand, care should be taken in specifying the information needs that this not only entails what the user wants, but also what the monitoring system is capable of producing[92, 248, 260].

2.3 MODELS FOR LINKAGES BETWEEN INFORMATION ON DIFFERENT WATER MANAGEMENT LEVELS

Monitoring often takes place at different administrative levels. Such networks at different levels need to be tuned to avoid duplication of work. Two models are presented here to characterize the logic of national and regional monitoring networks are presented: the pyramid model and the overlapping circles model. The pyramid model takes the position that at the lowest (local) level the most basic information is needed. This information can be aggregated to the desired information to be used on higher levels (regional, national) (Fig. 2.2). Essentially this is what is suggested in the hierarchical, iterative planning framework of Perry and Vanderklein[20]. All information collection in this model takes place on the lowest level. The information is therefore always an aggregation of basic information. The information needs on 'higher', more abstract levels will have to be supported by the 'lower' levels. The pyramid model as such supports the hierarchical, iterative planning framework. Also the steering model of Musters and others[138] can be applied, where the overall system represents the highest steering level while the lower levels are represented by the subsystems.

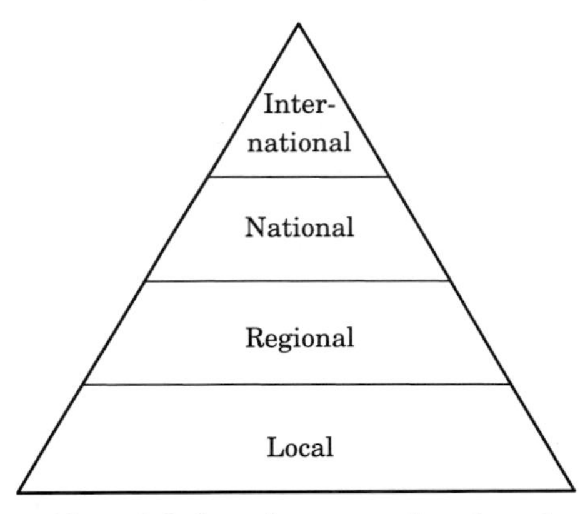

Figure 2.2 Pyramid model for coherence of various levels of monitoring

The circles in the overlapping circles model represent the information needs on the various levels (Fig. 2.3). Each level collects its own information and consequently supports its own information needs. In case of overlap, communication between the levels can take care of prevention of double work.

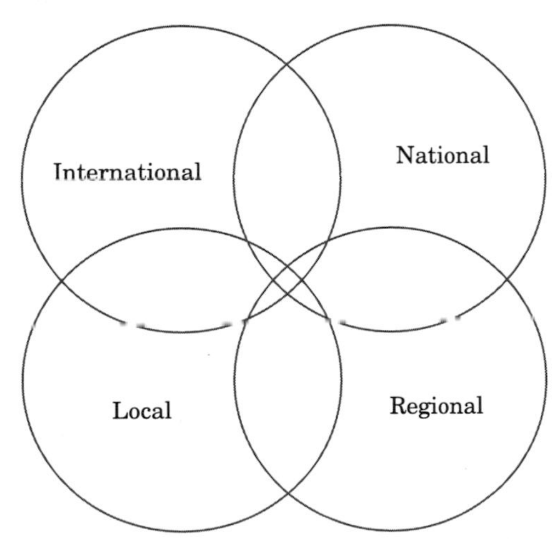

Figure 2.3 Overlapping circles model for coherence of various levels of monitoring

The overlapping circles model implies that the monitoring network can be developed independently from other water management levels. The information needs can be specified without taking account of other levels whatever system is present. In practice, the overlapping circles model exists in most situations. Different administrations collect the information they feel is necessary. When overlap is obvious, coordination takes place.

2.4 IMPROVING THE INFORMATION PRODUCT

Unfortunately, too often the information presented as a result of the actions as described in Section 2.2 does not satisfy the information users (Box 2.3). There are several reasons why this is the case[264]:

- Specification of information need has been insufficient;
- The information need as specified is not the 'real' information need (too little effort went into the process of defining the information need);
- The strategy of collecting information did not produce the right information;
- The information obtained generates new questions making the originally agreed upon information appear to be inadequate;
- The situation has changed (for instance new policies), causing other information to be needed;
- New methods have been developed, that demand different information.

Box 2.3 Asking information users about their information needs

Asking information users about their information needs often leads to bad results. Bemelmans[104] gives the following reasons for these bad results:

- Information needs are diverse and of a complex nature;
- Interaction between information users and information producers is weak;
- Some information users refuse co-operation for whatever reason;
- Information users are limited in their possibilities to specify information needs. This is related to the fact that people:
  - Have a tendency to refer to existing situations and are not able to come loose from it;
  - Put more emphasis on recently emerged information needs, although this does not cover for long term needs;
  - Have difficulty in detecting cause-effect relationships and consequently ask for the wrong information;
- Often make wrong judgments about risks and consequently ask too much information (one never knows what it may be useful for).

Generally, the resulting conclusion is that other information is needed. The new (i.e. changed) information need must be determined and the process starts again, but from a new position. Because the previous iteration has led to new, improved insights, this next cycle should lead to better-tailored information. By continuous evaluation and feedback, the cycle becomes a spiral, leading to a constantly improving quality of the information[41] (see Fig. 2.4).

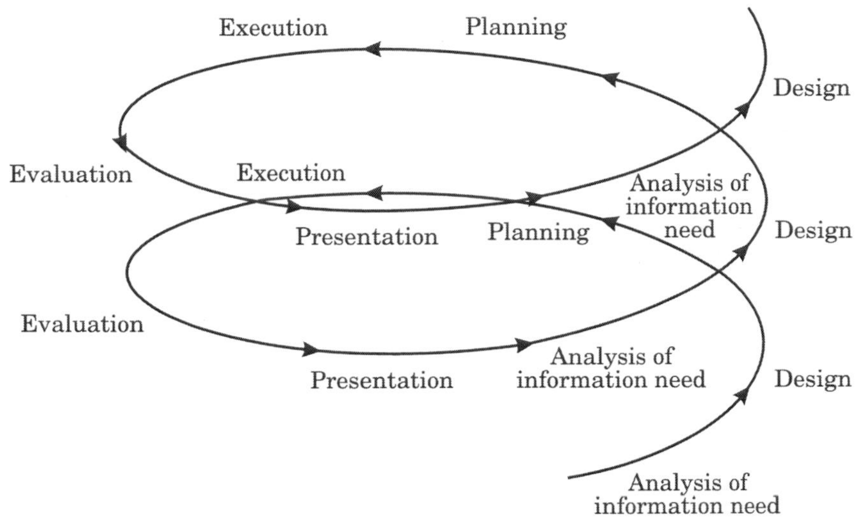

Figure 2.4 A quality spiral for the monitoring process[178]

The information cycle as presented in this chapter provides a conceptual model for information production that provides possibilities to manage the quality of the process. As the information cycle starts by specifying information needs linked to water management, it is based on water management goals and objectives. How this is exactly put into practice is the subject of this book. The cyclic character comprises regular evaluation of the gathered information. In this way, the information system design and operations are quantitatively connected with the information expectations and/or products required by management.

The way the resulting information will be used, influences the content and way of collecting information. Elements in the information cycle may put requirements on other elements. If, for example, a trend with a defined reliability has to be calculated, sampling has to take place with a certain frequency depending on the variability of the data. Calculation of loads of a certain parameter in a river requires data on concentrations of that parameter as well as data on the discharge of that river. Thus data analysis puts requirements on data collection. Similar dependencies exist for the other steps in the cycle; each step puts requirements on the previous step and limits the following step.

By theoretically going through the information cycle both clockwise and counter-clockwise, formulating the prerequisites and restrictions of every element, these requirements and limitations can be made explicit.

The information cycle offers a model that acts as a basis for the dialogue between information users and information producers. This is because the model is easy to comprehend and gives the information user a certain level of understanding of the information production process. It provides a method to direct the dialogue. With this, it is easier to come to those questions that must be answered to formulate an information strategy and that is accountable for the information produced. It can even be used to analyze the organizational set-up around information production in an organization. This can be done by identifying which departments in an organization or which organizations are responsible for what elements of the monitoring cycle. The logic in the information cycle should be reflected in the logic of the organizational set-up of the process. The information cycle and its monitoring version have been introduced in the Dutch water management bodies in 1997 and have become a natural and inevitable part of the common language since[266-268].

The information cycle largely views the water management process as a black box. It describes the order of activities to come to suitable information. It does not describe the flow of information through the water management process, as Gooch[216] correctly notes. Nevertheless, through the process of specifying information needs, the information cycle links the water management process to the information production process. The produced information links back to the water management process and decision system through the process of information utilization.

The information cycle is developed as a generic framework that supports several types of information collection, ranging from monitoring to modeling[269-271]. It also supports defining inventories and surveys. The nature of the resulting information production process is largely determined by the first step of the cycle, specification of information needs.

The information cycle in principle only cares about the process and does not deal with, for instance, institutional matters. In practice, this does not hold; as water management has to deal with these aspects, so has the information. The major activity where these issues come up is the specification of information needs. The structure of the information cycle provides support to the institutional structuring of information production as stated before.

The first step in the information cycle is the specification of information needs. The need for information of the information users is in this step translated into an overview of aspects that can be measured and that can be used to develop an information network. This is the essential step to link information production to the water management process and is extensively discussed in this book. The other steps in the information cycle will be shortly described below, but are not the focus of this book. Several other books exist that deal with these steps in more detail.

2.4.1 Information Strategy

The second step in the information cycle is the elaboration of an information strategy. The information strategy is needed to determine what part of the information needed will be collected from what source. Different sources of information as well as different ways of collecting and analyzing data are incorporated into the information cycle to provide integrated information (Box 2.4). Usually monitoring is used in water management to collect data. Models are often used to make predictions on the basis of the monitoring data. Surveys are used to provide insight into specific issues. Other sources of information may be data from other disciplines such as agriculture, recreation, sociology, ecology and economics, and other organizations like statistical bureaus. These different sources are then integrated in the data analysis step.

Box 2.4 Redesigning a monitoring program

Ongley and Ordoñez[249] describe an example of redesigning a monitoring program that uses different strategies that together comply with the information users' needs. The new program has four components: a primary network that provides long-term descriptive information on the status of important or sensitive water bodies (a combination of a variable-oriented and a risk-based strategy), a flexible secondary network that can respond to regulatory needs (regulation-driven), the use of surveys and specials studies to fill in the knowledge base for issue-specific concerns (investigative strategy), and a mobile emergency response capability for accidents (a combination of a risk-based and a predictive strategy).

Chapter 6 will provide more information on different information strategies, based on the information needs. Choosing a strategy may lead to rethinking of the information need. This is a feedback loop that can be addressed in a following iteration of the information cycle. It can also be addressed directly, turning back to the step of specification of information needs. In this way, it can lead to information needs that are adapted to this new insight directly. In the end this ideally leads to optimal information. However, one always has to bear in mind that resources are limited and, as a consequence, not every wish can be effectuated.

2.4.2 Data Collection

On the basis of the information needs and the strategy chosen, the data collection is conducted. Data collection in general comprises the setting up of a sampling network, carrying out of the sampling, analysis of samples and data storage. This is done in water quality monitoring. It can nevertheless also be used for the description of collecting, for instance, social data through interviews. This comprises selecting the people to be interviewed, doing the interviews, analyzing the individual interviews and storage of the results. Collecting data from other organizations entails describing what type of information is wanted. This requires discussions with that particular organization on what data they can produce and how this relates to the needed information.

Choice of sampling frequencies and locations is important in designing a sampling network. When for instance water quality data are available, use of statistics can minimize the number of locations through correlation between stations[272]. Knowledge of chemical, biological and physical processes, of specific local characteristics or properties of the object of investigation, and of the analytical methodology and statistics should also be used in devising a program of measurements[273]. Sampling has to be done in a manner that does not affect the outcomes of the analysis. For instance, water sampling for analysis of metals should not be done with the use of a metal bucket. Many of such sampling requirements are described in national and international standards.

Also laboratory analysis should be done with the use of standards and quality assurance. A quality assured laboratory is compelled to document its policies, systems, programs, procedures and instructions to the extent necessary to assure the quality of the results. It must also establish and maintain procedures to control (approval and review) all documents. The Quality Management System (QMS) should be based on the institution's quality policy statement to ensure managerial commitment. Elements of such QMS include properly maintained and calibrated equipment, the use of reference materials to calibrate methods, etc. More information on quality control and quality assurance can be found in literature[239].

Use of standards is especially important in transboundary monitoring to make data comparable. The data resulting from the laboratory analysis has to be stored in a way that enables easy retrieval, to avoid the development of 'data-graveyards'. Also, the data have to be documented in a comprehensible manner. Data should be validated or approved before they are made accessible to any user or entered in any data archive[274]. This includes validation of data. That means analyzing the data set to ensure that errors and inconsistencies in the data are identified and properly used in the further analysis. Errors that may occur are typing errors, outliers, detection limits that are too high, wrong units of measurement, mixing up of samples, and errors in sampling, storage, transport and analytical methods. These considerations and more are extensively discussed in literature[120, 128, 248]. Furthermore, clear agreements have to be made about the data that are stored. This includes information about sampling station, about the sample (location, date and time, sampling method, sampling depth, etc.), and the measurements results (analytical method, actual result including the unit, and an indication of the reliability of the result).

2.4.3 Data Analysis

Data analysis includes data handling, data processing and reporting. After storing the data, the data analysis process starts. Data are retrieved from their storage and, if necessary, linked to other data that may originate from other storages (data handling). Processing of these data is done to calculate for instance mean and median values, minimum and maximum values, standard deviation, trend, loads, etc. Many different statistical methods exist to calculate all the statistical data mentioned before (Box 2.5). The preferred method should be carefully implemented to ensure that the outcome reflects the intended analysis. The final step in this procedure is to report the statistical data in an agreed format like a table, graph, or otherwise.

Box 2.5 Limitations to the use of statistical methods[128]

"It probably goes without saying that statistical methods should be appropriate for the level of knowledge and experience of the data analyst and that the results produced should match the experience level of the information user. With modern statistical computing capabilities, however, there is a strong temptation to try out lots of data analysis methods without really understanding them. While understanding all the theory is not important, it is essential to understand the most important assumptions of a given method, whether the assumptions are reasonable, and the consequences of violations of these assumptions."

Ward[275] propagates the Data Analysis Protocol (DAP) similar to the sampling and sample analysis protocols to ensure that the outcome of the data analysis process is comparable over time and between organizations. The data analysis protocol includes components like: a statement of the exact information to be produced directly related to the specified information need; procedures for preparing a raw data record for graphical and statistical analysis; means to visually summarize the behavior of the water quality variables; recommended statistical methods which yield the desired information; and reporting formats for the resulting information.

2.4.4 Information Utilization

Decision makers and policy makers have to be given the opportunity to understand the implications of their decisions. One of the major constraints to stable, long-term water resources management however is not lack of knowledge, but transfer of that knowledge into a clear and understandable form to policy makers and decision makers. To remedy the poor transfer to policy makers of adequate knowledge and understanding, scientists should learn to produce clear and transparent statements[12]. Reports from the data analysis step are input for the step of information utilization where the results are translated into information that is relevant for water management; information users must be reached to convey the resulting message. The information product should therefore be relevant, accessible, and attractive for the information users.

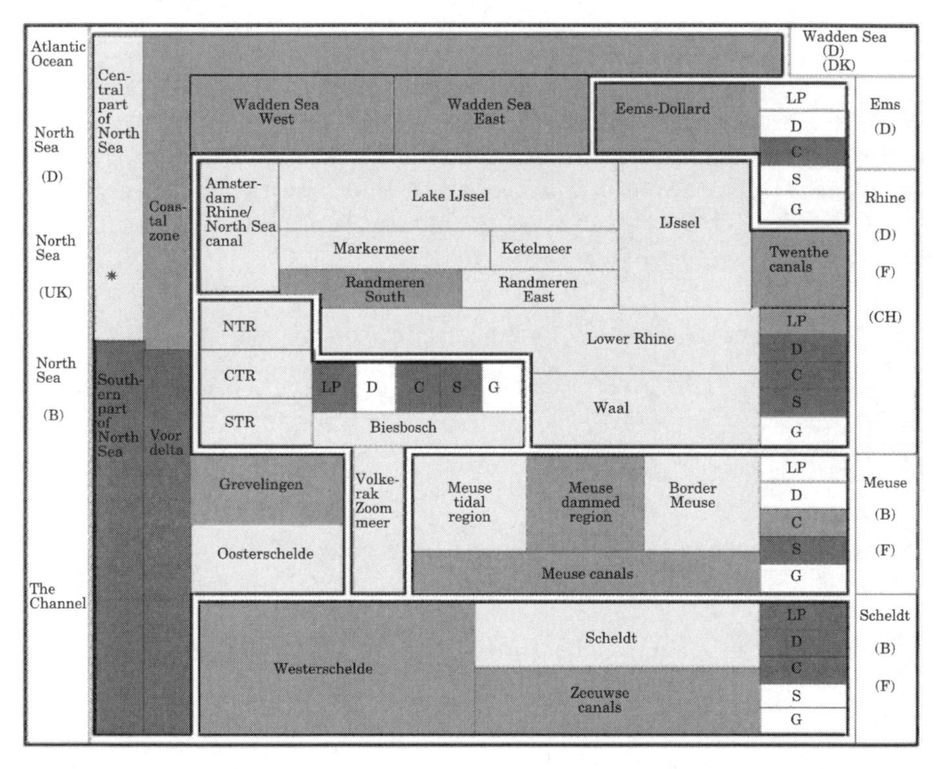

Figure 2.5 In the water dialogue software system, a schematic representation of the map of The Netherlands is presented called the Water Mondriaan. This graphical presentation shows the major water bodies. The color of each box represents the state of that water body[276].

Color image of this figure appears in the color plate section at the end of the book.

One of the important challenges is the aggregation of the huge amount of available data into clear information. To illustrate the volume of data, for each location on the Rhine River alone, the Dutch national water quality monitoring network collects some 5000 data points on water quality each year. Various tools have been developed for assessment and reporting of water quality data. For example, The Netherlands has developed the STOWA method for ecological judgment[277], the water dialogue[276] (Fig. 2.5), and the AMOEBE[278] (Fig. 2.6). In New Zealand, a presentation method has been developed for analyzing water quality trends[279]. Such tools are generally developed to make the information accessible and attractive. The information included in these tools is, however, not explicitly based on an assessment of the information users' needs. The resulting information is as a consequence not always relevant and therewith the usefulness and resulting application of the tools varies.

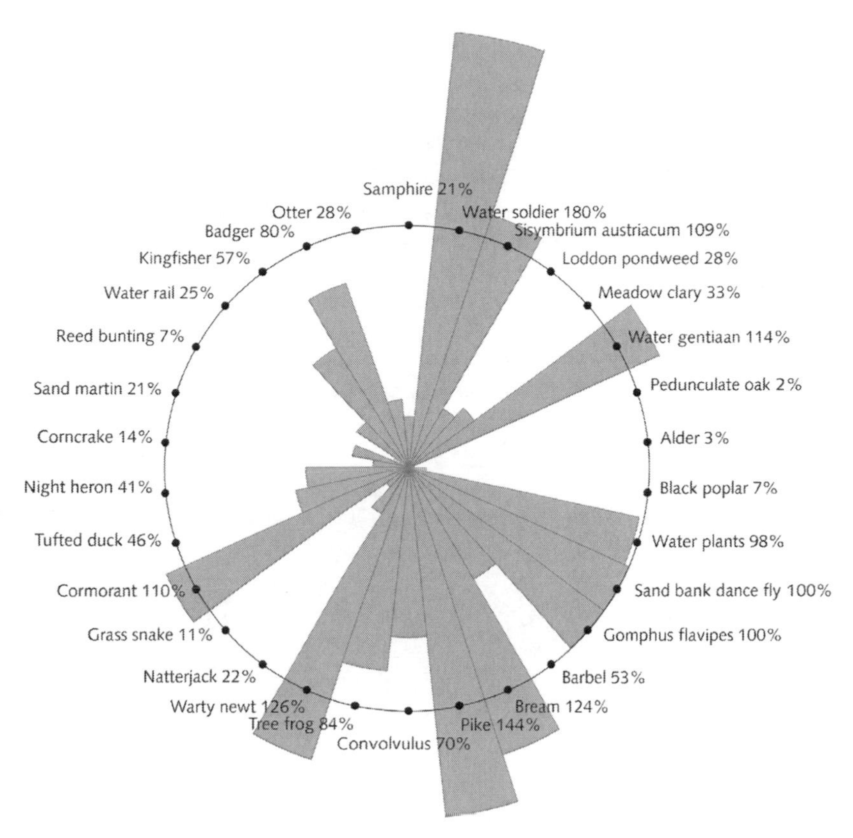

Figure 2.6 The AMOEBE show a series of indicators in a circle, in this case a series of animal and plant species. The circle shows the desired value, the segments show the actual situation. Ideally, the circle would be filled[278].

Denisov and Christoffersen[37] looked at the impact of information on decision making (Fig. 2.7). They state that decisions are not only made on the basis of considerations of individuals and institutions, but they are strongly influenced by visible and hidden systems of interests. The role of information is to help promote, develop and establish more formal management frameworks. These are supposed to modify the behavior of people or organizations in the desired direction, such as laws or economic mechanisms. There are several steps to take before information will have impact. The produced information is initially put down in the form of maps, graphics, books, etc. The way this information is subsequently communicated, for instance, through Internet, mass media or conferences, is important to reach a specific target group like decision makers or the public at large. This target group can develop ideas based on this information that in turn lead to changes in for instance, laws or policies, or even values. These changes then lead to changes in behavior, which finally result in a better quality of the environment. But however perfect the quality of the information is, and however perfect the information is presented, it also needs to be communicated to reach a maximum impact. Active communication yields more effect than passive reporting. Therefore, like in the step of specifying information needs, also in the step of information utilization close interaction is needed between information producers and information users. However, too active communication may become pushy, resulting in less impact[37].

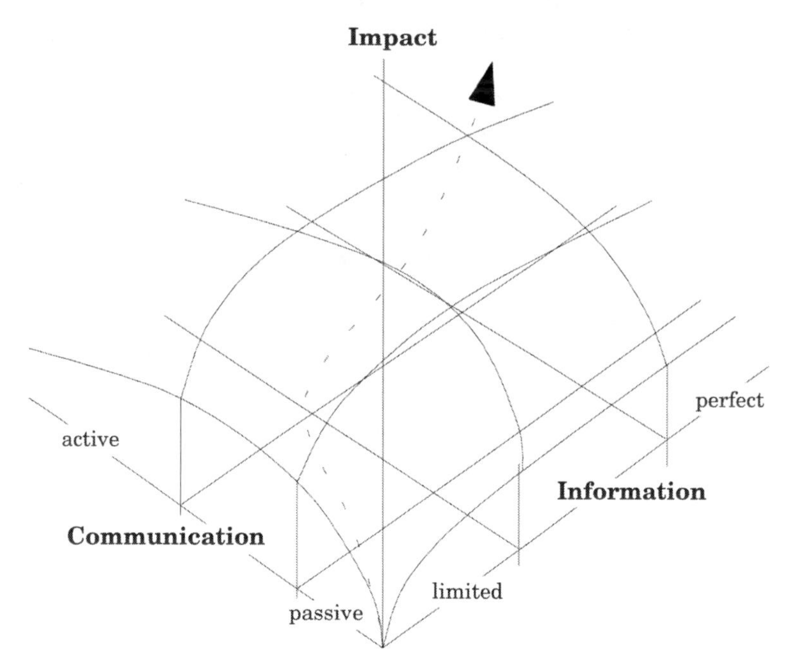

Figure 2.7 Impact as a function of the quality of information and a type of communication[37]

2.5 APPLICATION OF THE INFORMATION CYCLE

The information cycle describes subsequent activities in the process of producing information, but does not go into detail on these activities.

The cycle does not address issues like institutional aspects, allocation of responsibilities, quality control, and quality assurance, which are part of the overall process management. Regarding quality control and quality assurance, the information cycle can nevertheless be a basis for defining the distinguished activities, the contents, and the outcomes. An example of this approach is developed in Timmerman and others[239] where the distinguished activities are described from a QA/QC point of view in which the output of one activity becomes input for the next activity. In this chain of activities, the input requirements are guiding preceding activities. It is important that in designing the subsequent steps, choices made are traceable. In case of deviation of the information produced from the information need as specified, the point where the deviation originates can then be traced back.

In a similar manner, institutional aspects can be incorporated. Responsibilities attached to activities are assigned to individuals or institutions. The information cycle provides a framework to make the assignments. The information cycle in this way also includes prioritization and design of information products in the specification of information needs.

Each of the elements of the information cycle has to be elaborated, depending on the required information and the strategy to collect this information. Monitoring, for instance, may require methods of data analysis different from modeling. In the Guidelines on Monitoring and Assessment of Transboundary Rivers[179, 280] the monitoring cycle was publicized as the outline to develop cost-effective, accountable monitoring, justified by value and use of information, starting from the river basin approach and providing an integrated assessment of the river water quality (Fig. 2.8). This monitoring cycle is an adaptation of the information cycle.

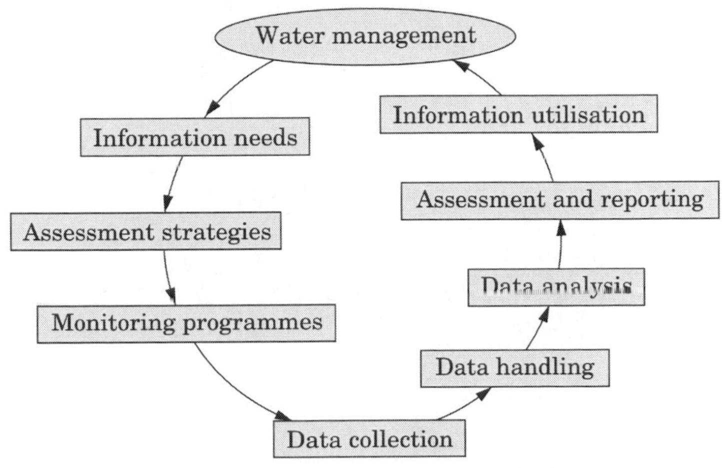

Figure 2.8 Monitoring cycle[179]

2.6 THE NWQMC MONITORING FRAMEWORK

The U.S. National Water Quality Monitoring Council (NWQMC), that was established in 1997, in its work to develop consistent and scientifically defensible water quality data and information, developed its own monitoring framework. This framework was intended to guide the activities of the NWQMC and its Methods and Data Comparability Board to facilitate communication among professionals and volunteers working in different aspects of monitoring, to guide the design of water quality monitoring programs, and to respond to the need for a warehouse of consistent information on water monitoring design methodologies.

The Council defined a framework as shown in Fig. 2.9. Before the flow of information can begin, the information goals must be defined on an operational level, along with a monitoring strategy designed to meet the goals. A monitoring design must be completed to guide operations involved in obtaining the desired information. Once operation of a monitoring system starts, the flow of water information begins at the interface between the water and monitoring system personnel. At the point when sufficient data are available to support analysis for an identified information goal, data analysis and interpretation, via graphical presentation, statistics, modeling, or some combination of these, takes place. Conveying information and results to information users may take many forms, depending upon the information need, timeliness sought, and the management style of the decision maker. The graphical representation of the framework includes six interconnected primary elements held together by the three C's (collaborate, communicate, coordinate)[281].

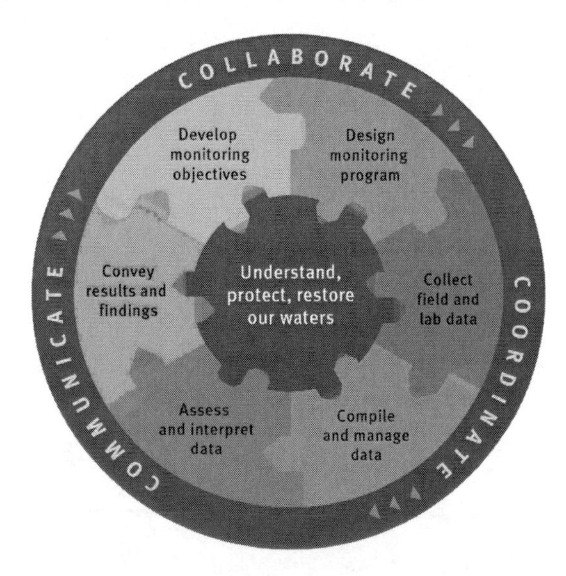

Figure 2.9 NWQMC framework for Water-Quality Monitoring Programs

When comparing the monitoring cycle and the U.S. NWQMC monitoring framework, the similarities in the various steps are clear. In the U.S. version the outer ring of shows collaboration, communication and cooperation while in the monitoring cycle these elements are included implicitly. The same goes for the arrows that in the monitoring cycle depict the sequence of activities. In the U.S. NWQMC framework, the arrows convey both the connection and the feedback deemed critical to smooth connections among monitoring components. As described, the feedback is implicit in the monitoring cycle. The monitoring cycle includes a box to emphasize water management as the key driving force behind efforts to harmonize monitoring concepts and terminology. The U.S. NWQMC monitoring framework appears to be more general in that it contains no reference to management. The center wording implies management is taking place, but it also permits the framework to be applied to monitoring conducted as part of a research program, completely separate from direct management information needs[281].

The NWQMC framework especially visually improves the monitoring cycle through including of feedback and blurring the borders between the activities. Also the outer and inner rings add to the improvements. The lacking of an explicit role for water management is however a reason to not adopt to this cycle but to stick to the information cycle (also see[282]).

2.7 EXERCISES

- In your organization, define what types of information are collected. Don't only look at monitoring information but consider this question in the light of the definition of the Aarhus Convention, i.e. including administrative measures, environmental agreements, policies, legislation, plans and programs (also see Box 1.5).
- How is monitoring organized in your organization and what types of information does this concern? Is/are there a specific department(s) that fulfill this task, how is the information collection organized and how is the information transferred to the information users?
- Try to relate the different steps in the information cycle to different departments of your organization and determine how the different departments communicate relative to how this is depicted in the information cycle. Also determine if the organizational structure complies with the procedural structure of the information cycle.
- What do the different processes within the steps of the information cycle look like in your organization? Is there a structured approach towards the information production, are standards used and what kind of standards are these (organization/national/international), are the processes described and if yes, in what form, is there QA/QC procedure, etc.?

- How is the overall process of information production managed in your organization? Is there an overall management of the process, is this described, is there a QA/QC procedure to ensure the whole process, are there procedures to ensure the links between the various steps, etc.?

3

How to Develop the Process

This chapter describes the rugby ball methodology for specification of information needs. The rugby ball signifies the initially diverging character of the process and the necessary convergence towards the end of the process. The chapter will first discuss various approaches for problem solving and the way these are incorporated in the rugby ball methodology. It then describes the rugby ball methodology providing a five step structure for the process that enables planning and communication, and supports selection of actors that need to be involved. After studying this chapter the reader would have obtained the following abilities:

- Understand what type of process the rugby ball is, also in view of the literature on this topic
- Understand the steps in the rugby ball
- Understand the purpose of each step in the rugby ball
- Be able to design a process on the basis of the rugby ball.

3.1 FRAMEWORKS TO MANAGE THE PROCESS

Strategic information should assist in structuring the issues at hand and clarify what decisions should be taken. Much of the issues are unstructured in policy making as discussed in Chapter 1. To deal with unstructured issues, 'strategic planning' was developed in the 1970s. The concept of strategic planning starts off from a structured approach; it aims at structuring the issues by working systematically towards making the targets more explicit and exploring and predicting the circumstances that influence achieving the targets. The structuring is followed by an assessment and evaluation of possible strategies.

In practice, this approach has some disadvantages in which the following arguments play a role[283]:

- Formalized planning is based too much on abstracted and aggregated data;
- Too little attention is paid to the strategic significance of concrete details and synthesis of findings;
- Managers need alternative views on strategic possibilities, not uniform schemes;
- 'Hard', factual data have limitations; at closer look they are not as 'hard' as they appear, they are not available in time and do not cover all relevant aspects of the problem.

The best solution according to Mintzberg and others[283] is a flexible approach. Different frameworks should be provided and the framework that fits best should be used for the specific situation. In other situations, other frameworks may be better applicable as will be discussed in Section 3.1.2. The preferred framework for a large part depends on the different ways different people think.

3.1.1 Different Ways of Thinking

Methods are developed on the basis of a way of thinking of people about reality and in turn are influenced by the different ways of thinking; the way people deal with decisions can be quite different. In general, two categories can be distinguished. First is a more analytical, systematic approach in which the decision maker aims at rational decision making. Second, is a more intuitive approach in which the decision maker decides on the basis of rules of thumb, intuition and experience. These two styles of thinking are distinguished as two outer ends of a scale. The one end of the scale is the design style that includes a recording style of observation and an inclination to solve problems in an analytical way. 'Design' people emphasize formal rules and explicit knowledge of the trouble spot like put down in handbooks. The other end of the scale is the development style where people are more inclined to observe in a sensing way, emphasizing on intuition and implicit knowledge of the trouble spot, which has more content than words can express. Emphasis is on developing ideas. These differences are further characterized in Table 3.1. The different types of thinking influence the effectiveness of methods and models used. 'Development' people will not easily adopt a rigorous method, while an implicit method will lead to agitation among 'design' people. Note that individuals will have a preference for one of the two ways of thinking. However, rigorous 'design' or 'development' people do not exist in reality. It is nevertheless important to realize which way of thinking is behind the method when developing or using a method.

Table 3.1 Characteristics of the design style and the development style[104, 283, 284]

Design style	*Development style*
The emphasis is on the formal organization: the organizational structures and the rules;	Emphasis is on describing the factual behavior of people in an organization;
The focus is on quantifiable aspects, decision should be based on solid data, not intuition;	Much attention for social and political, usually qualitative aspects;
The organization is considered to be a mechanism;	The organization is like an organism;
Separation between thinkers and practitioners;	Integration of thinkers and practitioners;
For changes, general design rules are used;	For changes, the specific situation is the starting point;
Changing is a linear, finite and discontinuous process;	Changing is a cyclic, open ended and continuous process;
People are part of the organization.	The organization is in the people.

Some other design-philosophies or -paradigms can be distinguished in information system development next to the design – development scale[104]:

- Objective versus inter-subjective reality: Some people assume that there is one (objective) reality that is perceived in the same way by everyone. Others feel that each person experiences reality in a different way, depending on knowledge and experience, role in an organization, etc.
- Rational versus semi-rational decision making: Decision making can be assumed to be rational, implying that decision procedures can be formalized. In practice, decision making can only be rational in well-structured situations; in unstructured situations decision making does not follow pre-set rules.
- Comprehensive versus partial approach: A comprehensive approach aims at one all-embracing overview, while the partial approach aims at decomposing reality into manageable parts.
- Deduction versus induction: The deductive school starts off from the more general, existing situation and tries to make specific improvements. The induction school moves from specific observations to broader generalizations and theories.
- Top down versus bottom up: This can be applied in different ways, for instance for the organization, starting from top management or the shop floor, or the design approach, starting with an overall general design or the specific rules, but in general it applies to the level of abstraction that is started from. Often, deduction is described as top down, while induction is considered bottom up.

- Long versus short lifespan: The information system can be designed to be unchanged for a period of at least 4 or 5 years, or is designed for the first year and will gradually be changed over time. Generally, the longer the planning process takes, the longer the expected lifespan will be.

These paradigms should be seen as the outer ends of scales as they are never totally implemented. Moreover, various strategies combine different paradigms into one approach.

3.1.2 Strategic Management Schools

Monitoring networks are very similar to tailor-made software systems in the sense that they are designed to be operational for multiple years. Development of such networks therefore requires strategic planning and management. As a result, different approaches towards strategic management will be briefly described.

The way management takes shape is largely dependent on the social, political, economic, and cultural context of the system to be managed[105], but also on the paradigms of the managers. The process of strategic planning generally involves determining the present position (Where are we now?), the new goals to be achieved (Where do we want to be?), and the process to reach these goals (How are we going to get there?[265]. Since the 1900's, many different ideas on strategic management have been brought to light that include these three questions. These ideas are however designed differently by different scholars, based on different paradigms. In their book "Strategy Safari", Mintzberg and others[283] distinguish three different (groups of) schools in strategic management; prescriptive schools (Box 3.1), descriptive schools (Box 3.2) and the configuration school. Their description of these schools will shortly be discussed in this section.

Box 3.1 Prescriptive schools[283]

Prescriptive schools are:

- The design school: This school appreciates strategic management as a deliberate, top-down process of planning, done by one (the CEO) or a few people. Ideally, the used model is simple and the strategy is tailored to individual cases.
- The planning school: This school takes the same steps as the design school, but uses formal planning processes, decomposed into distinct steps, supported by checklists and techniques. The CEO is responsible but staff planners do the execution.
- The positioning school: This school takes on the premise that there are a limited number of (generic) strategies, and the strategy formation process is one of selecting the best strategy, predominantly through calculations.

The prescriptive schools are a group of schools that are more concerned with how strategies should be formulated than with how they necessarily take form. These strategic planning schools propagate the analysis of organizations and its environment. Strategic planning can support strategy making in providing data input for this process, it can act as catalyst by providing planning schemes and techniques, and it can act as an evaluation tool to scrutinize the strategic plan. These schools can be placed at the 'design' side of the scale.

One of the important critiques that stress to be cautious of prescriptive goal-oriented decision making, also in strategic planning and management, is that it makes assumptions about the ability of policy makers and managers to control systems under their jurisdiction or responsibility[285]. Another critique is that decision making assumes both a steady state in the society and organization, and infinite environmental resources. Society, however, is a complex, adaptive system and should be treated as such[8].

The descriptive schools respond to this critique by considering specific aspects of strategy formation. They are concerned less with prescribing ideal strategies but with describing how strategies get made; strategy formation and implementation occur simultaneously. These schools can be placed in the 'development' type of thinking.

Box 3.2 Descriptive schools[283]

Descriptive schools are:

- The entrepreneurial school: This school relies heavily on the vision of the leader and as such starts off from the top-down paradigm. There is an active search for new opportunities, power is centralized, uncertainty provides opportunities for big leaps forward, and growth is the dominant goal.
- The cognitive school: This school sees strategy formation as a cognitive process that takes place in the mind of the strategist. A strategy is a concept that is not easy to achieve.
- The learning school: The complex and unpredictable nature of the organization's environment precludes deliberate control. Therefore, strategy making must be a process of learning over time and follows a bottom-up approach.
- The power school: This school propagates strategy formation as shaped by power and politics. Strategic maneuvering and shifting coalitions within and between organizations determine strategy making.
- The cultural school: In contradiction to the power school, the cultural school sees strategy formation as a process of social interaction, based on the beliefs and understandings shared by the members of an organization.
- The environmental school: In this school, the environment of an organization is the central actor in strategy making. The organization must respond to the forces coming from the environment and strategies must ensure proper adaptation by the organization.

The final school, the configuration school, is not really descriptive or prescriptive but takes an in-between position. This school combines all the other schools. Most of the time, organizations are more or less stable and during this period, particular strategies can be applied, depending on the characteristics of the organization. Between these longer periods of stability, there are short periods of transformation into another configuration. The configuration school could best be situated as a normative school in the sense that it provides guidance for the analysis and structure of the decision process by driving the thinking process in an organized way[129]. In this way it combines the best of both (descriptive and prescriptive) approaches. But as the configuration school refers to other strategic planning approaches, there is still the matter of what strategy to choose.

3.1.3 Choosing a Strategy

"Information strategy means making choices, not carving out a master plan in stone"[42]. This citation indicates the organic, ever-changing character of strategy. Davenport describes information strategy as having the following characteristics:

- Strategy is a continual, incremental process of setting and resetting organizational direction;
- Strategy should not be elaborate or detailed, because the future cannot be anticipated in detail;
- Strategy is a dialogue rather than a document;
- Business managers, not 'strategic planners', should do strategy and planning.

This is a 'development' type of statement that expresses what strategy is, but not how it should be developed. What would be the best way to go through this process of strategy development? From Section 3.1.2 it may be clear that there is not one best strategy; a strategy has to fit to the situation and needs of the organization that is performing strategic planning, and has to fit to the expectations of the decision makers in the organization. When choosing between these schools to find the best strategy, Mintzberg and others[283] pose some issues that may have to be addressed:

- How complex or how simple should the strategy be?
- Should the strategy be fully integrated or is a loosely coupled collection of strategies sufficient?
- Is the strategy unique or generic?
- Is the strategy formation process predetermined and controlled, or is it emergent as the process develops?
- Is the strategy formation process an activity of one individual or a collective process?

- Is the aim of the process to change the organization or to continue stability?
- Is it possible to make a choice or is each strategy bound by the possibilities of the outside world?
- How long can strategic thinking last and when should the action start?

For one thing, there is no generic solution; each of the schools has characteristics that make them useful in a certain context. Managing such a process is done best by picking the aspects that provide the best fit, both to the 'strategy maker' and to the organization, the information environment he or she is working in. The strategic management philosophies provide insight into ways to manage the process. Different approaches are available towards designing the process. The best approach is highly dependent on the dominant culture.

3.1.4 Soft Systems Methodology

Soft Systems Methodology (SSM) is a methodology developed by Peter Checkland[286]. This methodology was incited by the experienced inadequacy of systems engineering. The inadequacy was especially provoked by the assumption of systems engineering that is was dealing with well-defined systems that can be managed to achieve objectives, that can be unequivocally defined and that have a definite solution; the so-called hard problems. However, many problems are ill defined. That is that the objectives are unclear because of the complexity of the problem: the so-called soft problems. When the 'hard' system approach is applied to 'soft' problems, the result will only represent part of the requirements and as such remain unsatisfactory. This can be compared to the design (hard) approach versus the development (soft) approach.

SSM takes on the concept of a system as a set of activities of people taking action to deal with a problem situation. In general, this is the issue that is dealt with in this book. Traditional systems approach to problem solving is based on reductionism, solving one stage of the problem at a time. In unstructured and complex situations, this approach is not adequate. SSM provides an incremental, iterative approach that allows the human element of systems to be incorporated into system design, by structuring a debate[287-289].

SSM distinguishes between the real world and system thinking. SSM considers the different views of people by assuming that each individual will see the world differently[104]. This will often lead to varying understandings and evaluations of situations. The culture and politics of an organization will inevitably include diverse views that may not necessarily be opposed to each other, but may be different enough to cause problems in defining clear objectives. Because the process assumes people will have different views, the goal is to achieve consensual action by moving towards understanding of the varying perceptions. The practitioners of SSM must therefore be open to other people's ideas for the process to be successful[290]. For this reason, one specific element of SSM is drawing a rich picture. Rich pictures are a graphical representation of the understanding of the problem situation[290] showing the elements of slow-to-change structure and elements of constantly-changing process[288]. The rich picture is part of creating imagery of the situation and is very supportive in communication between the participants[291].

The basic idea in SSM is to formulate and structure the real world situation in a meaningful (from the systems point of view) way. An important tool to achieve this is the CATWOE analysis. The CATWOE stands for[290, 292]:

- Customer—The immediate beneficiaries (everyone who gains benefit from the system) or victims (if the system involves sacrifices).
- Actors—The people who do the activities as defined in the system.
- Transformation (process)—What the event may achieve; the conversion of input to output.
- Weltanschauung—The view of the world that makes the transformation process meaningful. Basically the Weltanschauung can be considered the definition of the problem that is dealt with.
- Owner—Proprietor who can ultimately direct the system and could close it down or stop it.
- Environment constraints—The external environmental constraints that limit what can be done, like organizational policies, and legal and ethical matters.

These six elements make up for a well-formulated root definition. Especially the Weltanschauung is important; this is directly related to understanding the problem and is also closely linked to the concept of the mindframe.

The stream of cultural analysis is also introduced in the SSM thinking to clarify the CATWOE analysis. The cultural analysis has three other types of analyses to be considered in order to structure the problem situation; an intervention analysis, a social system analysis, and a political system analysis. These analyses are implemented in the first stage. This stage involves getting more information about the system politics and power structure. The intervention analysis involves defining who the client is (who causes the intervention to take place), the problem solver (who conducts the study), and the problem owner (who is directly involved in the problem situation). The social system analysis involves defining the roles of each individual involved, the expected behaviors of each role, and the values to evaluate the performance of each individual. The political analysis finally involves defining the structure of power and the implicit organizational beliefs. This involves thinking about what makes an individual powerful within an organization and what are the symbols of this power[290]. The cultural analysis is complementary to the CATWOE analysis and helps to complete it.

SSM is a structured approach for clarifying objectives in complex, dynamic problem situations and as such fits with the requirements that are developed in this book so far. It is a goal driven approach, because it focuses on a desirable system and how to reach that. The process is geared towards dealing with people issues because it supports the analysis of unique perspectives on a problem situation. Especially the concept of the Weltanschauung links well to the concept of the mindframe as used in this book. SSM has a philosophy of continual improvement. The various stages can be iterated within the structuring process if improvement is needed. There is a hope for convergence of ideas of practitioners during the process. The information cycle, as described in Chapter 2, works in the same way.

The process is in some ways more important than the outcome. This is because the process will give the members the opportunity to learn more about the diversity of views about or within their organization, and also about their colleagues. It likewise helps to develop a vision and improve communication between people in the organization. The participative nature and strong focus on human activity systems of this methodology facilitates the development and testing of a systems model of a 'messy', poorly defined and complex problem area[287, 288, 290].

3.1.5 The Preferred Strategy

There is no one best way to approach strategic planning, as has been described in the previous sections. Different paradigms exist and based on these paradigms, different schools of thought can be distinguished. Each school has properties and conditions that make their specific approach more suitable. Another context however can make another approach more suitable. The preferred strategy is consequently depending on the situation. This is both dependent on the relation of an organization with its environment as was discussed in this section. And it also holds true for the way the organization itself deals with, for instance, information.

An interesting approach in this respect is the approach of Argelo and Boterman[293], who developed a method for information planning, Their method comprises four elements: concept, method, techniques and tools. The framework they use basically consists of three phases; 1) Develop a plan for the problem-solving project; 2) Elaborate on the problem situation; and 3) Develop a plan for the implementation of the solutions. Their approach is output-driven in this respect that their method focuses on the resulting products. They use various underlying concepts depending on the situation at hand. For the different activities in the process, various techniques and tools are used. The overall framework they use is not intended to be a linear model that has to be passed through from left to right. It is intended to be like a hopscotch diagram, where each product can be realized more or less independent from the others. As such, it is a configuration strategy, combining the best two worlds.

The rugby ball methodology as described in Section 3.2 builds on the approach of Argelo and Boterman[293]. The rugby ball has loosely connected elements that together result in a program of action. Products are described for each of the elements and methods are proposed to perform the different activities. Depending on the purpose and nature of the process, specific elements of the procedure will require more attention.

3.2 THE RUGBY BALL METHODOLOGY

The rugby ball methodology for specification of information needs adds to the hopscotch framework as developed by Argelo and Boterman[293]. The rugby ball adds interaction with the stakeholders to the first and second phase of developing a plan and elaborating on the problem situation as described above. This is done to ensure wider understanding, improved support for the process and its outcomes, and to avoid bias or overlooking of issues. In practice, to achieve this, two workshops are incorporated; one at the end of the first phase where the stakeholders can decide about the scope of the study and the actors to be involved, and one at the end of the second phase, where the stakeholders can decide if the right analysis is performed and if this will lead to an information network that meets their expectations. Through this approach, the stakeholders can minimize their effort, as they can largely focus on the workshops and maybe an interview. On the other hand it maximizes their input, as the process is clear and they can decide how their input will be included in that process. They can, for instance, decide to provide extra input next to the workshops. The methodology incorporates the notion of diverging in the first half of the process to enable input from various actors and to stimulate creative thinking, followed by converging towards a concrete outcome in the end of the process.

Based on these considerations, the methodology is constructed as a five-phase plan where all phases are interrelated. The overall structure of the framework has taken the form of a rugby ball as is shown in Fig. 3.1. The process starts from the left with the Exploring phase, in which the scope of the study is determined. This scope is discussed with the stakeholders in a workshop setting; the Initiation phase. Then, the problem structuring takes place in the third phase: the Elaboration phase. The results of the problem structuring are discussed with the stakeholders, again in a workshop setting; the Conclusion phase. Finally the process is described and a comprehensive overview of the problem structuring is developed that enables building an information network; the Completion phase. The content of the individual phases will be described in detail in the following sections.

Similar to the hopscotch framework of Argelo and Boterman[293] it presents a logical order that can be used as a hopscotch or route-planner in which the elements are filled in to achieve the full problem description. Iterations are possible if necessary and also intermediate workshops can be included. The rugby ball provides a basis on which the process can be build.

The shape of the image of a rugby ball (Fig. 3.1) symbolizes the initial diverging character of the process that at a certain point must converge into a coherent plan[294]. Combined with the structure within the phases, the figure turns into a rugby ball. This metaphor signifies that playing the rugby ball requires practice. That practice can be acquired in an organization by repeating the process regularly. It also points at the need for experienced facilitation of the process. Moreover, the rules of the game should be set. For instance, there should be institutional support, the starting points of the process need to be clear, etc. The generic rules are set out in this book; the rules that are specific for the process in a specific situation are set out in the first phase of the process as will be explained below.

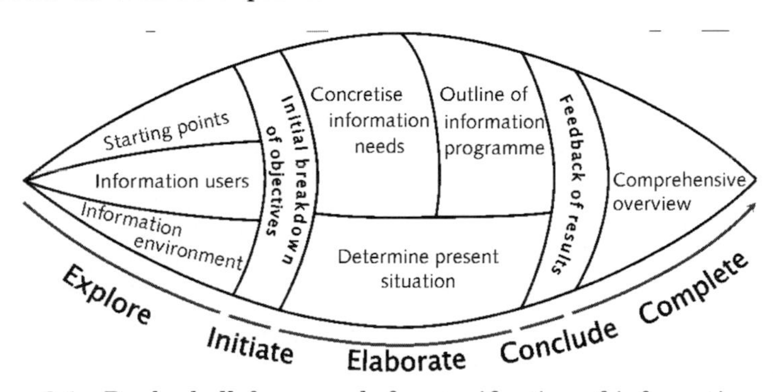

Figure 3.1 Rugby ball framework for specification of information needs

The representation of the framework as a rugby ball is a powerful tool in communicating the methodology. It provides an image that is easily remembered and recognized, which is helpful in the communication. It also helps to gain confidence as it is a solid image pointing at a solid process. It should be noted that the size of the elements in the figure does not reflect the amount of time or effort involved for the specific element but is merely designed in this way for a balanced figure.

3.2.1 Explore

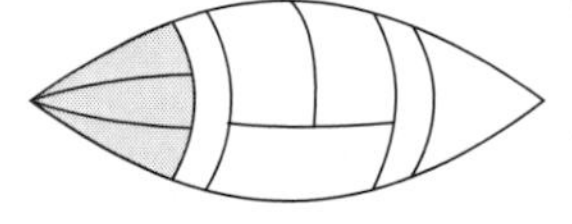

This first phase starts off with an intervention analysis, pointing out who is the initiator of the process to take place, who will conduct the

study (the 'problem-solver'), and who has a direct stake in the problem situation. Then, the boundaries of the project are established in this first phase by examining the starting points, by identifying the information users, and by inspecting the information environment. This is complemented by a social system analysis (what are the roles and values of the participants) and a political system analysis (what are the implicit organizational beliefs). The problem environment is clarified in this way. This phase compares to the CATWOE analysis as conducted in SSM, as will be described below.

The result of this first phase is an overview of the scope and goals of the information production. It provides an understanding of the situation under concern and a first structuring of the information problem situation. The first phase is largely conducted by the problem-solver, in close contact with the initiator (or owner) of the process. The initiator has to agree with the scope and goals before entering the second phase. Agreement on the outcome of this first phase with all the stakeholders, however, will be completed in the second phase.

The *starting points* are the first element in this phase to be elaborated. The starting points are examined to delineate the project at hand. Or in the words of Bardwell[231] "define the problem-space". This element includes the Transformation (T) and Weltanschauung (W) of the CATWOE-analysis. The essential question here is "What will be the subject of the study?" which relates to the Weltanschauung. In our case, the basic question is "on what part of the policy should information be produced?" It also includes looking at what elements of the natural and socio-economic system are taken into account.

Methods like boundary analysis or classification analysis[130] (Table 3.2) to describe what is included in the problem under scrutiny can be used in this phase[130, 295]. A subdivision of the problem can be made by relating a part of the entire problem to the categories subsystem, aspect-system or phase-system. Such distinction is helpful in the delineation of the problem as it not only provides handles to set the limits, but also points at the parts that will not be included. The subdivision in water quality monitoring is often a combination of two; choosing a subsystem like a sub-catchment, lake or river branch within the aspect-system of one or few policy objectives (Box 3.3). Karstens[296] provides an interesting example of how the choice of spatial and temporal scale influences the outcomes of a process. The subdivision is demarcated by policy aims or water management objectives as put down in policy documents or water management plans. This can also be a specific law which includes monitoring requirements, like the implementation of the Water Framework Directive in the national law of an EU Member State. The choice made here includes the selection of the conceptual models to be applied in organizing the problem as will be discussed in Chapter 5. A first analysis of the information needs is carried out in this phase.

Table 3.2 Methods of problem structuring[130]

Method	*Aim*	*Procedures*	*Source of knowledge*	*Performance criterion*	*Element in problem structuring*
Boundary analysis	Estimation of metaproblem boundaries	Saturation sampling, problem elicitation, and cumulation	Knowledge system	Correctness-in-limit	Metaproblem
Classificational analysis	Clarification of concepts	Logical division and classification of concepts	Individual analyst	Logical consistency	Problem sensing
Hierarchy analysis	Identification of possible, plausible, and actionable causes	Logical division and classification of causes	Individual analyst	Logical consistency	Problem situation
Synectics	Recognition of similarities among problems	Construction of personal, direct, symbolic, and fantasy analogies	Individual analyst or group	Plausibility of comparisons	Problem definition
Brainstorming	Generation of ideas, goals, and strategies	Idea generation and evaluation	Group	Consensus	Substansive problem
Multiple perspective analysis	Generation of insight	Joint use of technical, organizational, and personal perspectives	Group	Improved insight	Problem specification
Assumptional analysis	Creative synthesis of conflicting assumptions	Stakeholder identification, assumption, surfacing, challenging, pooling, and synthesis	Group	Conflict	Problem search
Argumentation mapping	Assumption assessment	Plausibility and importance rating and graphing	Group	Optimal plausibility and importance	Formal problem

Box 3.3 Sub-system, aspect-system and phase-system[105]

In a subsystem, only part of the set of objects is under study, like one specific human use or one river branch. In an aspect-system, only specific aspects of the whole set of objects are studied, like eutrophication that is relevant for many of the uses and functions of a water body. In a phase-system, only specific periods in time are considered, for instance the flooding situation.

Also, a water management analysis is performed. This includes looking at the relevant policies and legislation and at the problem situation. In a situation where the water management situation is not sufficiently laid down in policy documents, the water management analysis needed may be extensive. Chapter 4 describes how to perform a water management analysis. The function/issue table that is developed in the water management analysis is a good basis for discussion in the next phase of the process; the Initiation phase.

From this element it must become clear if the resulting information is used for instance for policy evaluation, policy preparation, operational purposes or maybe several of these purposes together. This is the Transformation analysis; it determines what part of the policy will be transformed into what type of information. The time scale during which the results must be valid should be determined in connection to this. For instance, it should be dealing with long-term policy or short-term operational management. This element of the Exploration phase thus provides a first clarification of the policy problem.

The *information users* form the second element in this phase to be decided upon. The selection of information users and therewith a large part of the actors to be involved in the project determines the outside world for the project. The question here is "Who will be using the information once it is produced?" In terms of the CATWOE-analysis this is the Customer (C). The group of information users involved represents the mindframes that will be included in the process. The general players in the integrated assessment that that should be involved, are[297]:

- The scientists. The fundamental role of the scientist is to know, describe and formulate the mechanisms of the cause-effect relationships, whether natural, technical or social. They are often the actors (A) in the process of specification of information needs but may also be customers (C). This will be discussed in the following section.
- The managers. The role of the manager (of for instance a monitoring network) is to help identify and structure the issues at hand, coordinate the whole process, communicate the approach and results, and implement the decisions made. The managers are often the owners (O) of the information producing process but can also be customers (C).

- The policy makers. The people to which the assessment is communicated in order to inform decision making. They are the owners of the policy problem and should demand a significant role in identification and framing of the issue. Usually they are the customers (C) in the information producing process but also have strong influences on the owners (O) of that process.
- The stakeholders. Individuals and groups that have a vested interest in the whole issue. They can often be identified as customers (C) of the process.

This listing implies that there is a wide variety of potential information users, ranging from directly involved policy makers to the public at large. Each of the information users will have goals and objectives that they wish to maximize[130] (It should be noted here that in a participatory approach that we are aiming at in this methodology, satisfying goals may lead to better results[184]). Inclusion of many information users leads to inclusion of a wider range of subjects into the study than originally might have been anticipated. A delineation of the information users is consequently needed to curb the scope of the study. The group of information users identified as relevant for the process should be involved at an early stage for them to be able to participate properly in the decision process. It is therefore important that this first phase in the framework is done in close consultation with representatives of the respective projected information user groups.

A balance however has to be found between inviting too many information users and too few information users. Inviting too many information users may lead to higher complexity of the process resulting in higher costs and prolonged consultation periods, to introduction of extraneous issues and the possibility of conflicts arising between different groups. Inviting too few information users may lead to exclusion or overlooking of some groups which in turn may result in resentment and non-cooperation among the information users[35].

The identification and selection of information users must be done in a systematic way and it must be recorded how this is done, because it clarifies the relationship between the content of the project and the process. It also clarifies the opinions of the initiators of the process about the various roles that the actors play in the process (for example user, producer, mediator). Inviting the right information users often depends on knowing who the people involved are[138]. To overcome this and to overcome overlooking of (groups of) users the information users as identified can be asked if they can identify any users they feel are missing.

The *information environment* is the third element to be dealt with in this phase. This element analyses how the information production process is organized. In terms of the CATWOE-analysis this element includes identifying the Actors (A) and the Environment constraints (E).

Actors here are the information staff; the people that determine whether all requirements for information production can be met. For this reason, they should be involved in the process. They are the monitoring people that will have to deal with the outcomes of the information needs specification process. By involving them in the process, the outcomes of the process are more likely to be useable and applicable to them. Unfortunately they are often mistaken for the Customers of the process.

The Environment constraints (E) entails a description of the relevant information processes; what type of information is processed how and what information is input for what process. These types of information and the linkages can be described with the use of for instance dataflow diagrams[298] as used in the information sciences. The relevant legal monitoring obligations as identified in the water management analysis (see Chapter 4) are part of this. Such obligations are based on national legislation as well as on bilateral and multilateral agreements. They form the Environment constraints in the sense that they set requirements on the monitoring. The monitoring budget can be another Environmnt constraint.

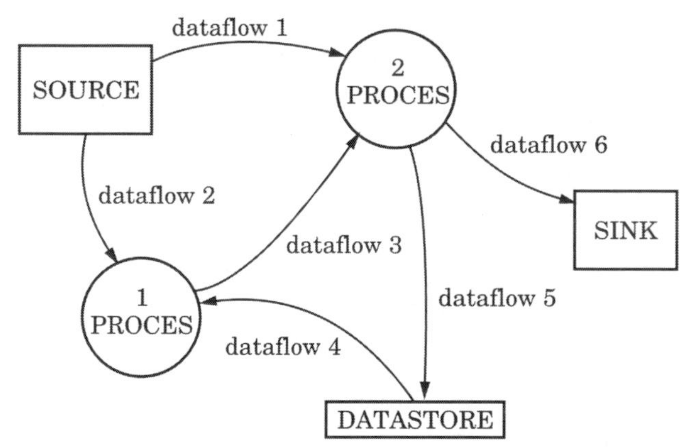

Figure 3.2 Example of a dataflow diagram

The Exploration phase thus forms the delineation of the project and the problem environment; what elements and what people are to be included, has to be clear after filling in the three elements in the Exploration phase. In effect, the CATWOE analysis is completed and the perspective, the actors, and the information environment in terms of the relevant processes and the relevant information for the project of specification of information needs are determined. These elements can alter in the run of the process. Fundamental changes in these elements however can only be realized through reconsideration of the project; that is when a new iteration of the information cycle is entered.

The result of the Exploration phase is the project basis. This basis should be agreed upon in the next phase, the Initiation phase, with the relevant actors to enable an effective participatory process.

3.2.2 Initiate

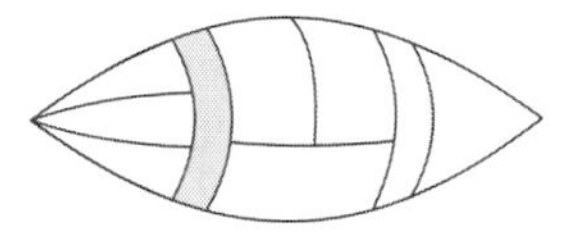

The goal of the Initiation phase is to bring all the participating people on the same level of understanding, make them agree upon the results of the previous phase, and to make them agree on the initial model of the information system. This requires a mutual appreciation among the different actors (information users and information producers) of each other's problems and leads to the necessary transparency and reduction of the subjectivity of the problem, as will be explained below. Emphasis in this phase is on communication and support of interaction and collaboration.

The results of the Exploration phase and a first analysis of information needs are discussed with the information users as a group. Input for this discussion ideally is a function/issue table and a first elaboration of the information needs based on the model of the information problem situation (see Chapter 5). An information needs hierarchy, as further discussed in Chapter 5, is a good method to present this first elaboration. A workshop is the preferred setting for the discussion, as close regular contact is needed between the different actors[41]. The workshop setting gives the opportunity to exchange ideas and intensify and consolidate mutual understanding. Identification of blind spots, discussing of conflicting ideas and further elucidation of assumptions takes place in such a workshop.

As the nature of the information problem situation varies from structured to unstructured, the strategy to cope with this situation varies[169]. The approach is quite simple for operational purposes, when the problem is well structured. The Initiation phase is essential but will become more complicated in a situation when the water management policy is not even clear and the problem situation is unstructured. An iterative approach will be needed in such a situation (see Table 3.3). The information users to involve vary, depending on the problem situation as well as the extent to which they are involved. Several methods and approaches exist to involve groups of people. An overview of communication techniques, their characteristics, target public and objectives are given in Jain and others[299]. A clear description of various ways of how stakeholders can be involved is provided by Ridder and others[300]. This book will not go further into detail on stakeholder involvement. The participation approaches as provided in literature can nonetheless be used when, for instance, designing the workshops.

Two methods are highlighted here that are fairly easy to apply and that provide good results; the Devil's Advocate method and Backcasting. Often, there is a tendency among the participants of a workshop to come to a quick agreement to be able to 'get back to work again' quickly[170]. If consensus is reached at an early stage, this is usually because differences of opinion are concealed instead of removed. This situation is not advantageous for specification on information needs,

Table 3.3 Strategies for the specification of information requirements[104]

Strategy	***Methods***	***Preferable situation***	***Level of uncertainty***
Questioning of information users	Interviews Questionnaire Structured interview	Well-structured problems, high level of knowledge and experience among information users	Low
Reference strategy: existing situation is reference	Comparison of situation with similar situations	Well-structured problems with temporal stability	|
Development strategy: thorough analysis and structuring of the problem	Critical-success-factor analysis Process- oriented methods	Moderately structured problems, moderate level of knowledge and experience among information users	|
Iterative or evolutionary strategy	Prototyping	Unstructured problems	High

which needs critical reflection. A confrontation of conflicting opinions is in such a situation very productive to reach a common goal. The participants of the workshop must be stimulated to have a fruitful confrontation of opinions. A method to reach confrontation is the Devil's Advocate method (Box 3.4). The method consists of appointing one of the participants as the devil's advocate. He/she is asked to critically question the statements made. This forces the other participants to argue why certain statements are used and if they are valid in the situation under scrutiny. Consensus is now reached through discussion, which strengthens the relevance of the outcomes. Application of the Devil's Advocate method in a process of specifying information needs is described in a study by Timmerman on assessing information needs for gradients[301]. The method helps to avoid the inappropriate use of cognitive shortcuts. An extra pitfall that is avoided through this method is a focus on alternative-focused thinking rather than on value-focused thinking[232] (Box 3.5).

Box 3.4 The Devil's Advocate method

The Devil's Advocate method is used to support strategic decision making. The goal of the method is to test if the line of reasoning on which specific policies or strategic plans are founded contains inaccurate assumptions or inconsistencies. The method in practice aims at postulating opinions as concrete as possible. Schwenk[302] describes the approach as inviting some employees to phrase strong criticism on the plan by casting reasonable doubt on as many assumptions as possible. The authors of the plan can react on this, with which the discussion is opened.

Another method that is described in the same case study on assessing information needs for gradients[301] and that is helpful in elaborating on the working scheme is Backcasting (Box 3.6). Participants in this method picture the situation that is to be reached within, for instance, 5 or 10 years. From this situation they work back to the present situation and try to determine what decisions would be necessary to reach the final situation and what information would be needed to take these decisions. This leads to an iterative process, looking back at the water management goals and objectives and rearranging the breakdown schemes (see Chapter 5). These two methods are useful in both the Initiation phase and the Conclusion phase. Brett[123] takes a similar approach when describing a methodology for specification of information needs where the first step is described as: 'Identify where we want to go, then work backwards'.

Box 3.5 Provocation as a method

De Bono[235] states that the main purpose of mind is to be brilliantly uncreative. He introduces lateral thinking, which is largely a way of taking a different angle of view towards problems, as a way to overcome this narrow-mindedness. He promotes the use of humor and provocation as means to invoke lateral thinking. A humorist often pushes the audience in a certain direction but with the punch line shows a totally different view of the situation through a double meaning of a word or a totally different situation than expected. Provocation does a similar thing by taking a standpoint contrary to the normal line of reasoning.

It is essential in the Initiation phase to convince the actors of the importance of their involvement. Experiences show that the best way to do this is to have the first analysis of the information needs done to such an extent that the gaps and points to be discussed at the workshop become clear. Several interviews are needed in the Exploration phase to identify the most important issues prior to inviting a wider number of people.

Box 3.6 Backcasting

"Backcasting is a methodology for planning under uncertain circumstances. In the context of sustainable development, it means to start planning from a description of the requirements that have to be met when society has successfully become sustainable, then the planning process proceeds by linking today with tomorrow in a strategic way: what shall we do today to get there? What are the economically most effective investments to make the society ecologically and socially attractive?"[303]. Backcasting is also used in development of scenarios. "Backcasting scenarios reason from a desired future situation and offer a number of different strategies to reach this situation"[95].

The selected actors are actively engaged in the Initiation phase. The Initiation phase aims at interaction between the actors and is preferably done in a workshop setting. The process is explained to the actors and the results of the Exploration phase are explained and subsequently discussed. The actors in this way become aware of the process and their roles in the process and are able to provide their input.

3.2.3 Elaborate

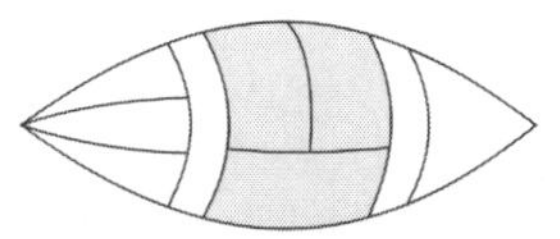

The actual problem structuring starts in this phase. After agreement on the scope of the study is reached in the Initiation phase, the problem structuring can start. The information needs as roughly specified in the previous phase is now further elaborated to enable development of an information network. Essentially, in this phase the transition from the real world to the system world takes place[290]. Most of this work is done by the problem-solver, who uses policy documents, legislation, monitoring network reports, and interviews and/or workshops to elaborate on the information needs, not in the least to fill in possible white spots and inconsistencies. The result of this phase in the methodology is what can be called the basic design[304] of an information network. The basic requirements of the information needs are at the end of this phase known and can be further elaborated.

This phase consists of three elements:

1. *Concretize information needs*. The information needs as specified and structured in the previous phase are further elaborated into the structured breakdown schemes as well as a first elaboration of the working scheme (see Chapter 5) is in this phase completed up to a level of detail that will in the end enable the design of an information strategy. This phase entails clarification of the problem from multiple perspectives and results in a clear translation of the policy issues into the information needs. Concrete indicators or determinands are specified as well as issues like response time (the time span within which the information should be available[305]) and the relevant margin (the information margin that is of concern to the information user. If, for example, the critical level of a pollutant lies at 10 mg/l, it is no use to provide information in µg/l. The relevant margin is about 2-5 mg/l in such a case[305]) (see Chapter 4) that are needed for network design. All this is done on the basis of the structured breakdown as described in Chapter 5. The different elements of the integrating decision model are elaborated here. Another aspect that needs attention at this stage is the information utilization. The information cycle is used as a tool to manage the entire process of information production and in this context is used both clockwise and

counter clockwise. This also involves looking at the use that will be made of the information after it is produced and then working backwards through the various phases in the information cycle to determine if what is needed in the end is supported by the information needs as the start of the process. Note that this is also a form of Backcasting. The use that will be made of the specific information is essential to determine the exact determinands and the specifics linked to them.

2. *Outline of the information program.* Based on the structured and detailed information needs, an information program is drafted. Essentially, in this element of the Elaboration phase, the working scheme is further elaborated. This element builds on the basic design and adds more specifics to the determinands including possible locations (or areas in case of ecological and morphological data, regions in case of economic or social data) and frequencies (again depending on the type of determinand this could range from continuous monitoring for measuring water levels to once in several years when measuring for instance perception of nature by local inhabitants). The result of this element should be sufficient for the design of an information strategy; that is the way in which the information will be produced (see Chapter 6). The resulting outline of the information program is also a basis for an estimation of resources to produce the information. This gives an indication of the necessary budgets and capacity needed for the production of the information. Moreover, it is also a basis for reconsideration of the information needs when the information cycle is closed and a new iteration starts (see Chapter 2). The conceptual model of the system is now fully developed
3. *Determine present situation.* An important aspect of this element of the methodology is to trace existing sources of information. Not all the required information will be available within the existing organization; other types of information are also needed. Moreover, much of the needed information will be produced in some form but may not be directly useful. The existing information networks should therefore be studied and compared to the information program. The relevant legal requirements as identified in the first phase should be included in the information needs here if this is not done already as a consequence of deciding about the policy objectives in the first two phases. If information, as collected in the already existing program, is not included in the information needs this can be because this information has become obsolete. It may also be an omission in the process of specifying information needs. Usually information networks are designed on the basis of implicit information needs, therefore the latter may well be

the case. To avoid overlooking certain (types of) information, the information that was collected under the existing network should also be taken into consideration for the future information needs.

The result of this phase should be a fully elaborated working scheme (see Chapter 5). As a large part of the work in this phase is done by relatively few people, in the phase that follows the results are discussed with the actors as identified in the Exploration phase. This ensures that the work was carried out according to the needs of the information users and producers, as will be described in the next section. Note that several iterations may be needed before the full set of information needs is finalized. And of course iterations of the information cycle will gradually improve these results.

3.2.4 Conclude

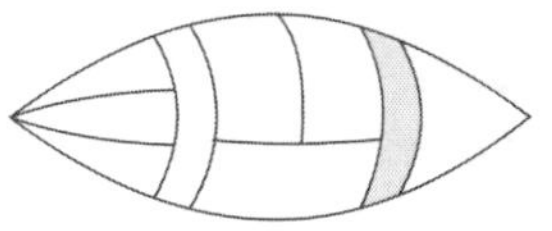

Emphasis in this phase is again on the interaction and collaboration between the various actors. All actors that have been involved in the process so far get feedback on the results and should now agree upon the draft information network as developed in the previous phases. Irrespective of the form of feedback, the people involved should be enabled to reflect on the results. Nevertheless, in principle the people of the first workshop meet again in a second workshop setting to discuss and adapt the results. As the level of detail in this phase can be rather high, many of the information users may be reluctant to join in. The workshop should therefore focus on filling the remaining gaps and solving controversies that came up, especially those coming out of interviews. The detailed information should nevertheless also be made available for the workshop participants.

The design of the workshop again depends on the participants and the issues to be dealt with. Methods like the Devil's Advocate method and Backcasting can be used to improve the outcomes of the workshop. Also other facilitation methods can be used. Note that the results of the workshop should be usable for its purpose; the workshop design must accommodate this. This will be further discussed in Chapter 4.

This phase is the finalization of the comparison of the conceptual model of the system to the real world. If this phase does not lead to agreement among the participants, depending on the level of disagreement one of the previous phases needs to be entered again.

3.2.5 Complete

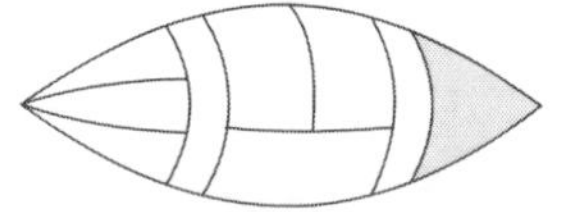

The results of the entire process and agreements of the Concluding workshop have to be documented. The result of this final phase is a comprehensive overview, a blueprint for the new information network, or a program of action. That will be the basis for the design of an information collection network. The result of the process is also needed to evaluate if the information as specified is produced in the end and if the right people were involved. This will be discussed in more detail in Chapter 6.

There are several reasons (like changes in policies and objectives or technical improvements) why information coming from the information network may not meet the information needs as defined as discussed in Chapter 2. The information cycle is, among others for this reason, a cyclic process and the process of specification of information needs will be entered again after some time. The form of the documentation, as developed in the process and finalized in this phase of the process, is put in must therefore be flexible and structured to enable easy changes and quick identification of consequences of changes. The structured breakdown schemes and working schemes are intended to support this flexibility.

3.3 EXERCISES

- What type of strategy would fit best in your organization; a design type of strategy, a development type of strategy, or an in-between strategy? What implications does this have for applying the rugby ball methodology in your organization? This could, for instance, imply that the rugby ball is presented as a fixed process structure or an overall image of the process to come.
- Position the rugby ball methodology in your organization; who is the owner of the process, who will perform the process, who should be involved in the process? Answering this question is the very first step towards starting the process. In your response to this question make use of the answers that you provided in the exercises in Chapter 2.
- What would be the first step to take when you actually want to start a process to specify the information needs in your organization? For instance, ensure commitment of the management, getting support from information users, other… The answer to this question could also be based on your assessment of what the need for the process is; if you assess that a serious 'water information gap' exists in your organization, the rugby ball process will be an important approach to solving this problem.
- What people would you try to involve in what step of the rugby ball process? Make a table of the five steps and the people that should become involved in each of these steps. Specify the role of each group in each step.
- From the above, develop a plan describing the steps you should take in a project to specify the information needs, the timing of

the various steps, and the milestones in the process for each of the steps in the rugby ball process.

- Identify the starting points; what are the policy objectives you should include in the process, what part of the water under the administration of your organization will be included, and what should be the outcome of the process?
- Describe for each step the information that you need in that step and how you will collect that information. This may include reports, policy documents, and monitoring plans, but also interviews or questionnaires.
- From the answers to the above questions develop a full CATWOE analysis; who are the Customers of the process, the Actors that will run the process, the Transformation from policy objectives to information needs (how and in what form), the Weltanschauung that describes the reasons for performing the process and the reasoning behind the choices made to run the process, the Owners or commissioners of the process that are responsible for funding of the process and possibly also for funding the outcomes (i.e. the information network), and the Environment constraints limiting the work to be done, for instance because of legal obligations.

4

Analyzing the Water Management Situation

This chapter provides a description of how to analyze the water management situation in a basin based on, among others, policy documents, inventories and existing legislation. Information needs are generally based on the water policy; the policy objectives determine what information is needed. The water management analysis as described in this chapter provides an overview of the water policy and the related elements. The analysis is needed to ensure that the proper elements are targeted in the information needs analysis. In some cases, the information is already available in the form of an elaborated policy document. The water management analysis in such cases can be limited to checking if all the necessary information is there; this chapter can in such a situation act as a check list. In other cases, much of the information is spread over several documents or institutions or may be not available at all. An extensive water management analysis is in such cases needed to make a comprehensive overview of the available information and it also enables identification of information gaps. Depending on the level of elaboration of the water policy and the policy objectives, the water management analysis will consequently need more or less attention.

After studying this chapter the reader will have obtained the following abilities:

- Understand why a water management analysis is needed
- Understand the different elements of the water management analysis
- Be able to develop a function/issue table.

4.1 THE NEED FOR A WATER MANAGEMENT ANALYSIS

The water management analysis provides the context of the water information system that is developed. It shows the present-day policy and the way it is translated into measures. It is the water management situation that the information producing system has to support through providing the proper information. The analysis is performed in the first phase of the rugby ball process (Exploration) as part of the starting points. It provides a baseline description of the characteristics of the water system under study, the relevant water policy and legal provisions, as well as studies and research reports on the existing problems. In many cases, this information is readily available in policy documents. The water management analysis in such cases is limited as reference can be made to the existing documents or a summary thereof. This chapter can in such cases be considered a checklist to assess if all the necessary background information is known. In other cases this information is not always available and an extensive inventory may be needed. The water management analysis in any case starts with collecting and classifying the available information.

Collection of the information entails, among others, the information on the geographical, hydrological, physico-chemical and ecological characteristics of the water system at hand. These characteristics provide a first impression of the situation from which the analysis can be built. The UNECE Second Assessment of Transboundary Rivers, Lakes and Groundwaters[306] for instance provides a baseline description of the transboundary rivers in Europe as well as the major water management issues in the basins. The descriptions in that report can be considered summaries of what the water management analysis can look like.

Other information that is collected entails, among others, information on uses of the water as well as land-use in the catchment area, problems that are encountered in water management in the water body and the receiving water body, legal obligations both in water management goals and in monitoring obligations, relevant existing classification schemes, norms and standards, and any further water policies and plans. All this will be discussed in more detail in this chapter.

The water management analysis is part of the Exploration phase of the process as described in Chapter 3. It sets the starting points for the process, delineating the 'problem space'; the part of the catchment and/or the range of policy objectives that will be included in the study. Figure 4.1 shows the various elements of the water management analysis that will be discussed in this chapter, the relationships between the elements, and how they are combined into a comprehensive overview of the water management issues.

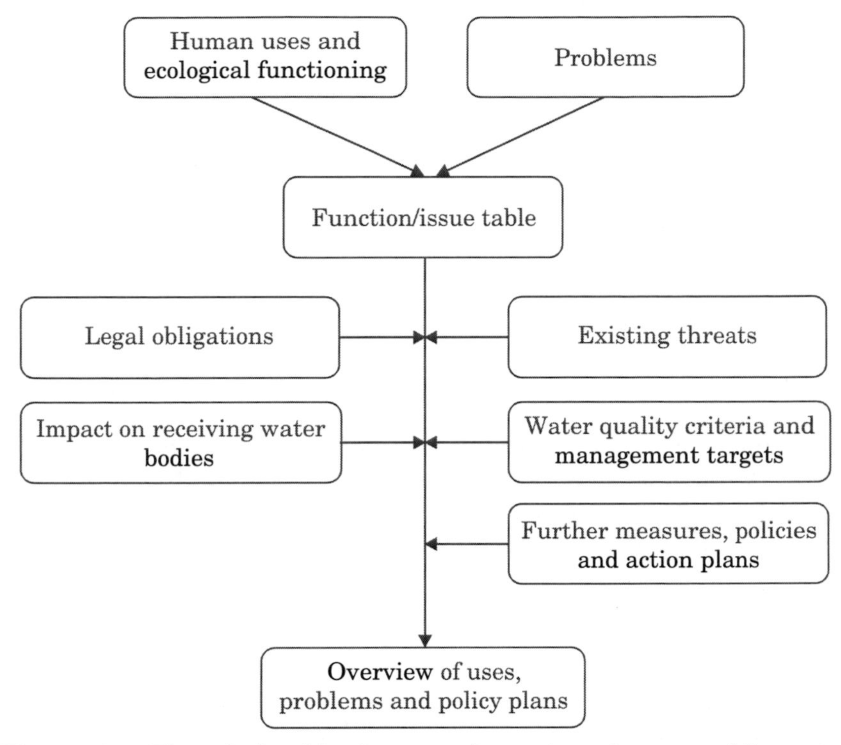

Figure 4.1 The relationships between the various elements of the water management analysis.

4.2 HUMAN USES AND ECOLOGICAL FUNCTIONING

Often, a wide range of human activities are concentrated in the catchment area of a river basin or aquifer. All these activities can have an influence on the functioning of the water body. Also the water itself is used for different purposes. Next to that, the water has a function in itself, that is called the ecological or ecosystem functioning. General uses and functions of the water are the following:

- drinking water supply
- industrial use
- water for irrigation
- fishing and recreation
- ecosystem functioning
- navigation
- waste water discharge
- land drainage
- mining (gravel and sand)

Profound insight of the situation around the basin is created by describing each of these uses in detail. This includes, for instance, the number of households that get their drinking water from the river, the type of industries and the amounts of water abstracted for process water or products, the type of agriculture and the amounts of water abstracted for irrigation, the amounts of waste water discharged by industries and households, etc. in this way, insight is built into the uses of the water and the potential pressures of these uses on the quality and quantity of the water. Waste water discharge, for instance, puts pressure on the quality of the water. If the amount of waste water discharged in a water body contributes a major share of the total volume of water, it is likely that there is a substantial pressure on the water quality. In turn, drinking water abstraction, industrial use, and fishing and recreation usually put high demands on the quality of the water. Ecosystem functioning requires good water quality and sufficient quantity. Irrigation water is somewhat less demanding on the organic quality but often requires larger quantities of water.

These examples show that the inventory of the water uses clarifies the potential problems for water management and preliminary indications of possible water management measures become clear. If, for instance, the same body of water is used for drinking water production and for discharge of waste water, it is obvious that high requirements should be put on the waste water discharge to ensure sufficient quality for drinking water production purposes. Moreover, it is obvious that information on the volumes of waste water, the quality of the waste water, the volumes of water abstractions for drinking water production, and the quality of the abstracted water is needed to keep track of the situation. Similar reasoning is possible for the other uses of the water. If the water management measures are added to this analysis, information on these measures is added to the inventory. For instance, the percentage of waste water that receives extra treatment in the case of a water policy of improving waste water treatment or the amount spend on investments into additional treatment facilities provides information on the level of implementation of the policy.

By prioritizing the different uses and functions, insight is obtained in the minimum characteristics that the water has to live up to, i.e. the minimum amount and quality of water needed for the specified uses and functions. For instance, water used for drinking water production puts high requirements on the water quality while water used for cooling purposes puts requirements on the refreshing rate to avoid temperature to become too high. Prioritizing of the uses and functions is done on the basis of policy documents and interviews with decision makers responsible for water management. Priority is usually based on social and economic factors like the number of households depending on the drinking water source or the economic value of the industrial production.

4.3 PROBLEMS

A second step is to make an inventory of the problems as encountered in the basin. Problems are recognized as such if they in one way or another hinder reaching the water management objectives connected to the use of the water and the ecological functioning. Problems range from water quality problems of a diverse nature, hydrological problems, and problems with landscape structure like river regulation and erosion. It is important to distinguish between the different problems, for instance the different types of pollution, as each type of problem may need a different response. General water quality and quantity management (present and future) problems and possible sources of the problems include:

- organic pollution: municipal waste water, industry (for instance sugar mills)
- eutrophication: agriculture (fertilizers), municipal waste water
- dangerous substances pollution: atmospheric deposition, industrial discharges and leakages, agriculture (pesticides, drugs), municipal waste water (detergents, drugs)
- accidental pollution: industrial accidents, dump sites (flooding, leakage), damages to oil pipelines
- changes of ecosystem structure and biodiversity; river regulation: navigation, ecological compartmentalization (migratory fish) by dams and weirs, loss of habitat
- risk of flood events with risks to the population and infrastructure
- water scarcity: irrigation, water-abstractions for industry, land degradation
- risk of contamination by flooding: through dump sites and contaminated sediments
- risk of erosion: land use
- saltwater intrusion and salinization
- bacterial pollution: municipal waste water
- high concentration of nitrates in ground water
- polluting substances having impact on the receiving river/lake/sea
- land subsidence and sea-level rise

All these problems have a source; this can be a specific pollution source or diffuse (atmospheric) pollution, but it can also be land-use change that imposes problems on the water system or the pressure from abstraction of water out of the water system. Responses to tackle a problem are usually most effective if they are targeted at the sources of the problem. Important reasons why the use of the water and other human activities that influence water becomes a water management problem are:

- insufficient methods of sewage collection and treatment in cities and towns
- lack of sewage systems and sewage treatment plants, especially in rural areas
- pollution discharged into soil and infiltrating ground water
- agricultural diffuse pollution caused by inappropriate storage and application of manure, fertilizer and pesticides
- infrastructural works like river training, constructions of dams and reservoirs, etc.
- climate change leading to hydrological extremes
- water abstractions and exploitation of ground water resources
- discharge of cooling water, causing a rise in temperature of the water

Describing each of these problems in detail provides deeper insight into the nature of the problem and the uses it affects. The description includes for instance the amounts of pollutants used (fertilizers, pesticides) or produced (amount and type of waste water), or includes ranges of river discharges in case of flooding or water scarce situations. It also includes estimation of the effects, either expressed in qualitative or quantitative terms. Regarding accidental pollution, estimates can be made of the impacts of industrial accidents, flooding of dumpsites, and other possible pollution sources. From this information, an assessment can be made of the impacts of problems on the receiving water and its uses and functions. Now, by listing the existing uses and functions, and the related problems, an overview of the actual water management situation is created. This overview is summarized in a function/issue table.

4.4 FUNCTION/ISSUE TABLE

The function/issue table is a method for adding context to the variety of water management objectives by linking the functions or uses of the water body to issues and problems. This enables framing of a diversity of interpretations into a common frame[151], thus simplifying the situation[152]. It is also a way of limiting the information[231] as it allows focusing on one function-issue combination at a time after the table is constructed. The table provides a basis to elaborate on objectives, as for each function/issue combination the objectives must be determined. A function/issue table (Table 4.1) is constructed based on the analysis of the problems and opportunities for the water body, with uses and functions on the horizontal axis and the problems and issues on the vertical axis.

The example of the function/issue table in Table 4.1 shows how the functions and issues on a national or sub-national level can be linked and classified. Next to that, the example shows how issues can be classified as a common concern of different institutions or countries. This is important in a transboundary context where the table can help to develop a common understanding of the water management situation between the riparian countries. In a situation where the water system under study is under the jurisdiction of different administrations, also on the national level, the table can thus help to find common ground for the development of an information system, as the example shows.

Even when the quality of the water system is no problem for the uses of that water, it may cause problems for the water body in which it discharges, a river, a lake, a ground water body or a sea. A description of this impact should be added as a problem for the water body as a whole. Impacts range from practically none for a very small contribution in volume, to substantial contributions, for instance, in terms of loading with nutrients and other pollutants. Table 4.1 shows how such impacts on receiving water bodies can be included in the table.

Table 4.1 Relations between the functions of the River Bug basin, the utilization of the water and the problems occurring in the basin (adapted from[307])

Problems	***Uses/functions***							
	Ecological function	***Supply of drinking water***	***Agriculture***	***Fish-farms***	***Recreation and angling***	***Supplies for the industry***	***Transport medium including sewage***	***Impact on lake***
Pollution by nutrients and eutrophication	+++	+++		+	+++	+	+	+++
Microbiological pollution	+	++	+	+	+++		+	+++
Organic pollution	+++	++		+	+		+	+++
Accidental pollution	+	+	+	+	+		+	+
High variability of flows	+				+		+	
Flood hazard	+	+		+	+	+	+	+
River regulation, damming and draining	++				+		+	
Pollution by toxic substances	+	+	+		+		+	+

Highly important (dark shading) | Moderately important (light shading) | Not important as common concern (no shading)

+++ High stress ++ Medium stress + Moderate stress

The function/issue table is developed in the Exploration phase (Chapter 3) to delineate the problem space and can be used in the Initiation phase as input for the initial discussion among the actors. The table is primarily targeted to support elaboration of the second fill-in question of the integrating decision model (see Chapter 5) and may not cover all relevant objectives. For some functions and uses of a water body for example, objectives may exist that are not directly linked to one of the issues. The objectives derived from the table must therefore be complemented with additional objectives, which were not included from the table according to the structured breakdown scheme. The full set of objectives is the basis for further elaboration of the information needs.

4.5 LEGAL OBLIGATIONS FOR MONITORING

Environmental legislation and agreements often include monitoring obligations. These obligations can be considered preconditions for the monitoring; they usually prescribe what needs to be monitored. In practice however, not all legal obligations are fully implemented[58]. The reasons can be that the necessary effort is too high in view of the available budget or one or few obligations may be just overlooked in the vast range of detailed obligations. In general monitoring should enable:

- Assessment of meeting the environmental quality standards which are defined by national and international law;
- Development of an environmental protection strategy at various levels of management;
- Observation of the progress in water quality related to water uses and the ecological policy of the state and at lower government levels;
- Assessment of the performance of legal instruments in the field of environmental protection, i.e. permits, assessment of influence on environment, spatial management plans;
- Ecological education and supplying information for the society;
- To ensure the information for the state administration;
- To ensure general water quality classification, qualitative trends assessment and prognosis.
- Monitoring obligations are laid down in:
- International regulations, like EU regulations (for instance[77, 308]);
- National regulations on surface water, waste water and ground water
- National regulations on the quality of water for specific uses, like drinking water or irrigation water.
- Bilateral Agreements

Listing of these legal requirements provides a full overview of the monitoring obligations. These can be compared later in the process with the information needs as translated from the policy objectives. By merging the two, a comprehensive overview of the information needs is made. The question that arises from the overview of the necessary information is how to collect all that information. This will be discussed in Chapter 6.

4.6 CRITERIA/TARGETS FOR FUNCTIONS/USES AND ISSUES

The next step is to develop an overview of the water policy. This includes inventorying the policies, measures and action plans as laid down in documents. The inventory also contains the water management targets that are mentioned in these documents. Plans are set out, for instance, to improve municipal waste water treatment through biological treatment, and by connecting households to Waste Water Treatment Plants (WWTP's). Also, plans may exist for revitalization of watercourses, introduction of best agriculture and foresting practices, rehabilitation of dangerous dump sites and introduction of clean production practices in the industry.

Next to inventorying the policies, an inventory of possible water classification systems and water quality criteria in the existing legislation is made. Often, legislation provides the legal basis for collection of information as well as the range of information elements that is considered useful, as stated above. Target values for quality of surface waters are often given for the following six categories:

1. organic pollution
2. eutrophication
3. acidification
4. toxicity
5. mineralization
6. bacterial pollution

Standards for these categories are set for, among others, use of water as drinking water source and water for irrigation. The water management targets, classification systems and criteria provide the information elements that are defined to assess the policy situation during the process of developing the water policy and the accompanying legal obligations (Box 4.1). They also provide information on the requirements of the information system. For instance, the water quality criteria determine the relevant margin of the chemical analyses; if the standard is set in mg/l, the analyses have to provide numbers in the range of that dimension, and consequently the information should reflect that dimension.

Box 4.1 Principles for water-quality objectives and criteria

Water- quality objectives and criteria shall[309]:

- Take into account the aim of maintaining and, where necessary, improving the existing water quality;
- Aim at the reduction of average pollution loads (in particular hazardous substances) to a certain degree within a certain period of time;
- Take into account specific water-quality requirements (raw water for drinking-water purposes, irrigation, etc.)
- Take into account specific requirements regarding sensitive and specially protected waters and their environment, for instance lakes and ground water resources;
- Be based on the application of ecological classification methods and chemical indices for the medium- and long-term review of water-quality maintenance and improvement;
- Take into account the degree to which objectives are reached and the additional protective measures, based on emission limits, which may be required in individual cases.

4.7 FURTHER MEASURES, POLICIES AND ACTION PLANS

Water management is a continuous process. This implicates that when management targets are set, thinking about new measures and plans has already started. Developments in water management that can occur but are not yet included in policy documents and/or legislation are for example:

- Improvements in waste water treatment including connection of households to waste water treatment plants, investments into new waste water treatment plants, and increasing the efficiency of waste water treatment in the existing plants.
- Building of household sewage treatment facilities when building of a collective sewage system is not economically and technically justified.
- Introduction of clean production practices in industry.
- Protection of water resources against pollution through promotion of ecological methods of farming and applying Best Agriculture and foresting practices. Development of emergency measurements to fight/control accidental pollution.
- Rehabilitation of dangerous dumps.
- Reduction of drinking water consumption by minimization of losses in waterworks and more reasonable handling by consumers.
- Transposition of existing legislation to EU legislation.

- Measures to support water retention and alleviate negative effects of flood events.
- Stricter control and reasonable utilization of water sources.
- Implement programs for the protection of ground water.
- Capacity building.

Awareness of such ongoing developments is important as it can influence the information needs; some of the information needs may become obsolete as a result of a development that may not have been included in the water management targets yet, but may already have had its effects on or influences the target setting.

4.8 OVERVIEW OF MANAGEMENT TARGETS

The information that is collected as described in the previous section is accumulated into a comprehensive overview. This section describes how this is done. In principle, this overview should cover all crosses marked in the function/issue table (see Section 4.4).

The results from the water management analysis, as summarized after a first inventory in the function/issue table and complemented with further analysis of the water management situation, are recapitulated. The table lists the most important management targets and links these to problems that may hinder reaching these targets (see an example in Table 4.2). Although this is not the intention, the table will show an inclination towards merely describing problems. However logical this tendency may be under the fact that the function/issue table targets problems, it should not lead to overlooking the fundamental objectives that are connected to the management targets. This risk of overlooking was an important reason

Table 4.2 Overview of management targets and the most important problems related to it

Management target (The management targets are related to the continued use and functioning of the river basin. Therefore, the uses and functions from Table 4.1 are reflected in this table. Added to the management targets are items that are mentioned in Sections 4.5 to 4.7)	Problems/Important aspects (The problems or important aspects to the management targets are a specification of the more general listing in Section 4.3. Basically, each cross in Table 4.1 is specified in this table, supplemented with items mentioned in Sections 4.5 to 4.7)
Drinking water supply must be secured for the future	The organic substances concentrations are threatening drinking water supply
	The N and P concentrations in the raw water are too high
	Concentrations of dangerous substances are too high
	Accidental pollution occurs

Table 4.2 (*Contd...*)

Table 4.2 (*Contd...*) Overview of management targets and the most important problems related to it

Water for industrial purposes must be available from the river	See drinking water. If conditions for the river water quality are in compliance with raw water for drinking water production, then industrial purposes are possible (If special standards are set for industrial uses, reference to drinking water should not be applied here)
River water can be used for irrigation	See industrial purposes (If special quality standards are set for irrigation, reference to industrial water should not be applied here)
Commercial fishing must be possible	River water quality does not comply with standards
	Sediment quality does not comply with standards, this may cause accumulation is fish tissues, making fish unsuited for consumption
	Bacteriological condition of water is unsuitable
	The morphological conditions of the river must be such that enough suitable habitat for commercial fish species is present
Recreational swimming in the water must be possible	Bacteriological situation of water is not in compliance with standards
	Water is unattractive for swimming
	River water and sediment quality are insufficient
The river ecosystem must be of good quality	The abundance of different species is not leveled, some species are dominating
	River water and sediment quality are insufficient
	The morphological conditions of the river are such that the development of aquatic communities not limited
Surplus water from precipitation must be carried off quick enough, but enough water must be retained for dry periods	Too little water is retained
Minimize impact on receiving water body	Concentrations of polluting substances in downstream region are too high

for developing the integrating decision model. The integrating decision model not only takes account of the problems and issues, but also looks at the fundamental objectives and the policy measures. The integrating decision model will be discussed in depth in Chapter 5.

4.9 SOME EXAMPLES OF DERIVING INFORMATION NEEDS

Already from the water management analysis, conclusions can be drawn about possible information needs. Using the above, two examples are presented here to demonstrate how information needs can roughly be derived in this way. Note that these examples do not provide a full inventory, but are merely included to show some logical paths to come to preliminary conclusions.

Example 1

Suppose a water management analysis in a Member State of the European Union in which:

- the inventory shows that a surface water is used or will be used for the preparation of drinking water;
- from the inventory in the water management analysis it is concluded that no Persistent Organic Pollutants (POP) such as organochlorine pesticides and polychlorinated biphenyls, are to be expected in meaningful concentrations;
- information from the upstream neighboring country shows that no POP are 'exported by the river' and that this will not change in the future;
- the Legislation Report shows that the EC directives 75/440/EC and 79/869/EC are applicable and are giving the most stringent restrictions for POP regarding this function/use of the surface water;
- additional surveys have not proved that POP are present in meaningful concentrations.

The conclusions are:

- the levels of POP are not important for the use and functioning of this surface water;
- this is not likely to change in the (near) future;
- a very low frequency of monitoring for POP can be applied.

According to European Commission (EC) directive 79/869 - dealing with the methods of measurements and frequency of sampling and analysis of surface water intended to be used for the preparation of drinking water - the minimum frequency for monitoring for POP under these circumstances is once per year. These data have to be reported to the European Commission in a prescribed format to be used for compliance testing of the EG-directive. No (further) measures are required to reduce the content of POP in that surface water, hence no additional necessity for monitoring exists.

Example 2

Suppose a water management analysis in a Member State of the European Union in which:

- the Inventory Report shows that a surface water is used or will be used for the preparation of drinking water;
- the Inventory Report shows that heavy metals are ubiquitous in discharged waste water, leading to meaningful concentrations in the surface water;
- information from the upstream neighboring country shows that heavy metals are 'exported by the river' and that this will not change in the near future;
- the Legislation Report shows that the EC directives 75/440/EC and 79/869/EC are applicable and are giving the most stringent restrictions for heavy metals regarding this function/use of the surface water;
- existing monitoring data and additional surveys have proved that heavy metals are present in meaningful concentrations, and that for some metals the limit values are exceeded;

The conclusions are:

- the contents of heavy metals are too high for the intended use of the surface water;
- actions to reduce the input of heavy metals are required.

As the input of heavy metals into the water is rather wide-spread, short term actions will hardly have any noticeable effect on the contents of heavy metals. Since levels of heavy metals are much too high, there is no need for very frequent and precise monitoring.

Together with the planning of the long term actions, appropriate information needs can be drafted, including verification of the effects of these plans. For instance, to establish a decrease in heavy metal contents at a certain sampling location of for instance 30 % over a 5 years period.

It can also be foreseen that by that time more precise monitoring is needed for compliance testing. EC directive 79/869 mentions a monitoring frequency of four times per year, and also the required precision, bias and detection limit of the methods of analysis.

For the European Commission a monitoring frequency of four times per year is enough. However, it may show that this frequency is too low to make a reliable estimation of the—30% trend. In that case, the latter information needs will define the required frequency of sampling and analysis.

4.10 EXERCISES

- For the situation you are working in, determine if the water management analysis will be an extensive analysis to collect all the necessary information or if most of the information is readily available. If the policies are well elaborated it may well be that most of the necessary information is at hand. Nevertheless, an analysis of the available documentation is needed to determine about this.
- Determine what elements you would include in the water management analysis. First do this by rote and then use this chapter as a checklist to reiterate.
- Will you be able to tell from the water management analysis where the information system under development will focus on? The first screening of the material can provide much information about the potential focus.
- Make a listing of the most important issues in the water system that you are working on and determine what functions and uses are affected and what the source is of the problems. This can either be done from the documentation or, if this does not exist or is outdated, publicly available information may bring you much of the necessary information. A simple inventory of the human activities that may influence the water system will already provide substantial information to answer this question.
- For each function-issue combination as determined in the previous exercise determine the relevant water policy and the relevant legislation, if available. This will enable to conclude about the desired situation and the actual situation.
- Identify the receiving water bodies and assess if these water bodies are affected by any situation in the water system that you are working on. The water system under scrutiny will drain in another water system that may be affected by the quality of quantity of the water drained.
- Determine the standards and norms that are applicable for the situation you are dealing with. National and international standards that apply for your situation are inventoried here.
- Make an overview of the relevant water management targets and the most important problem related to each target. The water

policy and possibly water management plans will usually provide this information.

- Reconsider the CATWOE analysis as performed in Chapter 3. On the basis of the results of the exercises in this chapter, the CATWOE analysis may reveal new insights.

5

Transforming Water Policy into Information Needs

The water management analysis as described in Chapter 4 is the basis for the transformation of the water policy objectives into information needs. The water management analysis provides the context of the water policy and the range of policy objectives on which information is needed. Now, a translation will be made of the policy objectives into the information requirements. To do this, a structured breakdown model is developed that provides a stepwise refinement of the information needs. The structured breakdown model is based on three existing conceptual models. The structured breakdown model can be modified by applying other conceptual models that may better fit the situation or are more familiar to the participants in the process, as will be described in this chapter. In substituting the models, care should be taken that replacing these models may lead to omissions in the information needs when the models applied do not cover all aspects as delineated by the integrating decision model.

After studying this chapter the reader will have obtained the following abilities:

- Understand the structured breakdown model
- Understand the different models in the structured breakdown scheme
- Be able to develop a structured breakdown of policy objectives
- Be able to develop an overview of information needs from the breakdown

5.1 THE INTEGRATING DECISION MODEL

Policy objectives are normally not defined in a way that enables direct translation into information needs, as discussed in Chapter 1. Decision makers have nevertheless much knowledge about the objectives and this tacit knowledge must be translated into explicit knowledge to derive information needs. This is a process of socialization and externalization of knowledge[193]. It can be performed through the rugby ball process as described in Chapter 3 (socialization) combined with focusing the attention of the actors on the decisions they have to take sooner or later, building on the information that needs to be available (externalization). The latter is described in this chapter. This chapter will show that by working systematically with the use of a few simple questions and schemes an almost full view of the information needs can be developed[129]. This works best, if it is based on a conceptual model or cognitive map of the decision situation[154].

The basic conceptual model that is used for the methodology is the model of the core-elements in water management[179]. The three elements; functions, problems, and measures, are in essence the elements on which the water manager needs information. These core-elements are the basis for the integrating decision model to structure the problem as shown in Fig. 5.1. The actors in the process of specification of information needs must reflect on the fundamental objectives that are

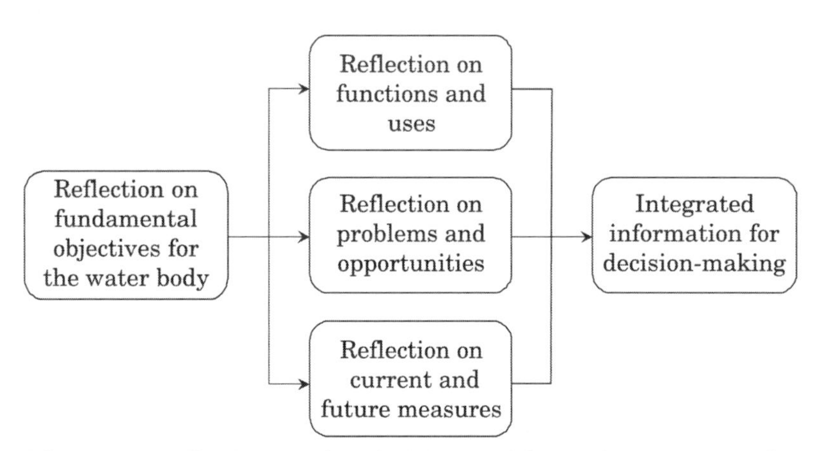

Figure 5.1 The integrating decision model as a basic concept for specification of information needs[301]

relevant for a water body[234]. They do this by thinking along three basic lines, formed by the three core elements in water management. This helps in pondering over the fundamental objectives and in developing an objectives hierarchy[310]. The three basic lines are:

1. What do I know or do I need to know about the functions and uses of the water system under scrutiny?
2. What do I know or do I need to know about the problems encountered and about opportunities for the functions and uses of the water system under scrutiny?
3. What do I know or do I need to know about the current and future water management measures?

The three lines of the integrating decision model together fill in the explanation function of monitoring as they enable to give insight into the relationships between the functions/uses, problems, and measures. The integrating decision model gives an overview of the current situation, the possibilities and constraints in reaching the objectives, and the current and future actions in implementing the policy decisions[301]. It comprises the four functions of monitoring; the accounting function by defining clear objectives, the compliance function through clear description of problems and opportunities, the auditing function by listing the measures, and the explanation function through the combination of the three[130]. The decision context is in this way clarified.

Defining this integrating decision model is the first delineation of the problem situation. The process of demarcation should be closely linked to the policy objectives as defined in the decision-making system. These are, in this situation, the policy objectives of the water management authority, who is the owner of the problem[292]. The link to the policy objectives ensures that the information can act as a constitutive force in the decision-making system[138, 205]. The integrating decision model is thus the starting point for the structured breakdown framework.

Filling in the three lines of the integrating decision model will be discussed in the following sections. The filling in will be done on the basis of three conceptual models, one for each line. Note that these conceptual models are chosen for a balanced overall framework. The framework can be modified by replacing the models as proposed by other models. These can be models that are more familiar to the participants in the process of specifying information needs or are better suited for the specific situation under scrutiny. Caution should be taken when replacing one or more of the conceptual models that the overall balance as defined through the integrating decision model is maintained.

Box 5.1 Objectives and attributes

'An objective is measured in terms of attributes'[237], or 'attributes are the criteria linked to objectives'[234]. Another term used in this context is 'descriptor'[311]. A descriptor is defined as the attribute or characteristic of an option that is the most adequate in the specific problem-context. Other terms used with a similar definition are measure of effectiveness, measure of performance, criterion, and evaluation measure[234, 311].

The structured breakdown schemes are further developed until the individual attributes of interest and determinands connected to that attributes are stated (Box 5.1). These attributes finally form the input for the working scheme. The structured breakdown approach thus makes the translation from the fundamental objectives into easily measurable or perceptible attributes along the three lines of the integrating decision model. The attributes have several properties to make them suitable; they must be relevant for the objective, understandable, measurable, and operational[294]. The set of attributes besides this must balance completeness and conciseness, and balance simplicity and complexity (Box 5.2). The selection of the attributes is an expression of the values or mindframes behind the fundamental objectives. The configuration that is built through the structured breakdown helps in generating imagery of the problem and in developing metacognition by dividing the problem into manageable parts[231]. The logical question that follows from filling this scheme is: 'If I have this information, do I know enough to be able to take a decision?'

Box 5.2 Different types of attributes

There are essentially three types of attributes: natural attributes, constructed attributes and proxy attributes[234]. Natural attributes have a common interpretation to everyone. If the objective is to minimize costs, a natural attribute is cost, measured in the local currency. In environmental management natural attributes are often too complex to use, like for instance the objective that the water quality should be non-toxic or pollution should be reduced. When it is impossible to measure all possible toxic effects or all polluting substances, a constructed attribute in the form of a subset of the natural attributes is used. This is often done through classification systems to assess the quality of a river or lake. A proxy attribute is an attribute that reflects the degree to which an associated objective is met but does not directly measure the objective[237]. The proxy attribute is valued for its perceived relationship to the achievement of the objective[234]. The use of proxy attributes typically reduces the number of attributes necessary for a decision situation and simplifies the description of consequences. This reduces the effort required to gather factual information while increasing the effort necessary to specify the value model. The motivation for building such a model is the same as for any model; namely, the intent is to have the model lend some insight into a complex situation to complement intuitive thinking[234]. In most practical cases the evaluation of alternatives has to be based on proxy attributes. Examples are pollution levels, water contamination, noise, and so on. With proxy attributes, data collection and organization is simplified[129].

The structured breakdown provides the link between the real world in terms of objectives and the systems or information world in terms of attributes (also see Box 5.3). The decision model underlying this process is made explicit and gives an understanding of the process of thinking, thus following the breakdown of objectives as promoted by Keeney[234]. Here the staving off of solutions is started, before the interests, needs and concerns are explored[231] and a meta-structure is developed through the process of problem search that results in the meta-problem[130].

Box 5.3 The information needs hierarchy

The 'information needs hierarchy', as a special form of a value tree[234], is a structure that is helpful to visualize the breakdown of objectives towards information needs. Going from general to very detailed, a hierarchy in objectives and related information needs is described. Starting with a general matter, through specifying different aspects of this concern more detailed (fundamental) objectives are elaborated in a top-down approach. Structuring of the hierarchy can also be done in a bottom-up approach, starting from available alternatives. Both approaches are included in the methodology. If building an information needs hierarchy is done in an iterative way, the general concern is broken down into the objectives, identification of factors of influence and measures or means objectives[122, 237, 294, 312]. The structured breakdown schemes are consequently simplified information needs hierarchies.

A paper by Bernstein and others[313] on planning tools to develop monitoring programs is used here to illustrate the development of an information needs hierarchy. They describe a hierarchy of different levels of information. 'Public concerns' are at the highest level, followed by 'Assessment issues', 'Policy goals', and 'Scientific and management monitoring objectives'. The 'Potential measurements' are at the lowest level. In their example they elaborate on the public concern: 'How safe is it to swim in the bay'? This is a rather broad, not yet fundamental objective. This public concern is translated into the assessment issue 'Health risks from swimming and surfing'. Now, it becomes clear that the safety issue is translated into health risks. The ensuing policy goals are to 'Protect the public from health risks associated with swimming in the bay' and 'Reduction of pollutant inputs to the bay'. Now the scientific and management monitoring objectives are stated: "Using a suite of effective microbial indicators sampled daily at shoreline stations throughout the bay, along with data on human contaminant inputs, ensure that public health standards are met, that illegal discharges are eliminated, and that information on swimming conditions is rapidly communicated to the public" . This can be broken down into two monitoring objectives, namely: 'Collect data on human contaminant inputs' and 'Use indicators to ensure that public health standards are met' Also two measures can be distinguished, namely: 'Illegal discharges are eliminated' and 'Information on swimming conditions is rapidly communicated to the public'. Bernstein and others[313] finally list three potential measurements, namely 'Bight-wide water quality', 'Beach warnings and closings', and 'Sewage spills'. The resulting information needs hierarchy could look like the hierarchy that is shown in Fig. 5.2. It will need further elaboration, however.

This stepwise focusing and reduction of the scope is in line with the strategies to build understanding of problems as depicted by Bardwell[231]. By using the structured breakdown frameworks, input from various disciplines can be included, thus supporting an interdisciplinary approach. The hierarchy visualizes the breakdown and is therefore a powerful tool to communicate with the actors. This latter point is also important as this methodology should support a learning or negotiation strategy[169] that in turn enables broadening of insight into other disciplines[125]. The information needs hierarchy also supports priority setting. By prioritizing issues on a high level, the priority determinands are derived. Prioritization helps in choosing what part of the information needs must be realized given the limitations in capacity and budgets[314].

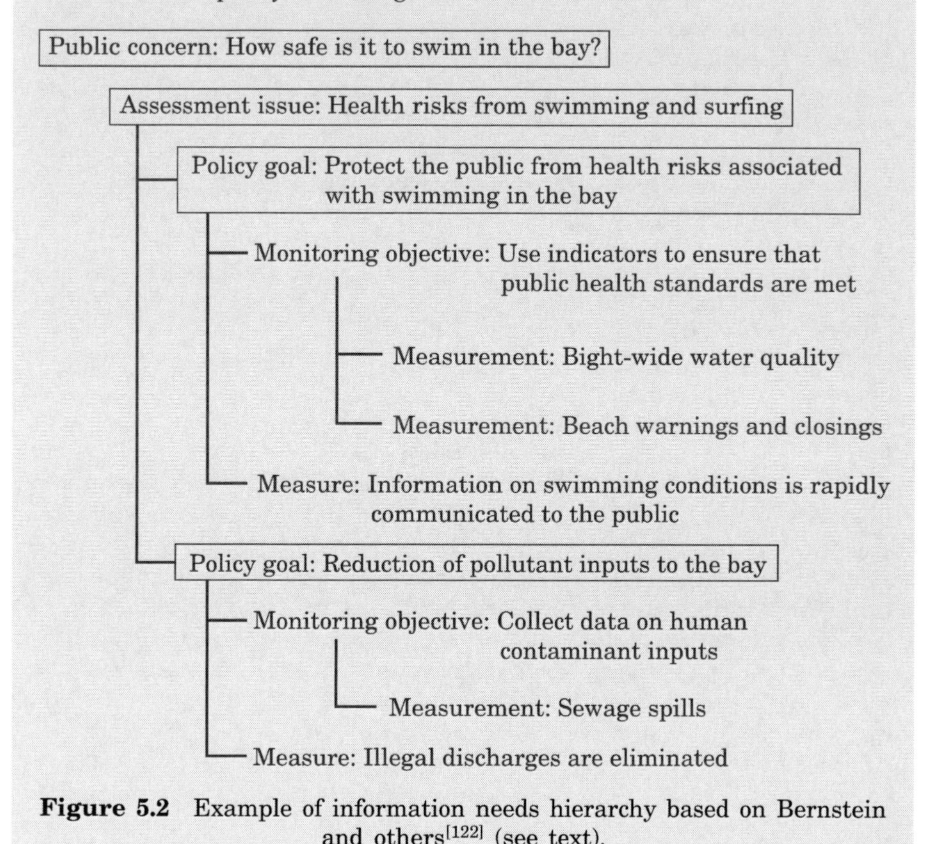

Figure 5.2 Example of information needs hierarchy based on Bernstein and others[122] (see text).

5.2 STRUCTURED BREAKDOWN OF FUNCTIONS AND USES

The first line in the integrating decision model is to clarify what information is needed about the functions and uses of the water system. This can be determined through the objectives for the functioning and uses of the water system, as specified in the water policy and legislation. The information is needed to determine to what extend the objectives are reached and over time also to verify if the situation improves. The latter relates to the accounting function of monitoring. The information user will know if the objective is achieved if information is available on all the attributes specified from the question 'From what do I judge if the objectives for the functions and uses are achieved?' The information user will be able to evaluate if the objective is likely to be achieved over time when this information is collected on a regular basis. The latter enables showing the changes over time. To find the attributes, it is useful to distinguish between the significant aspects of an objective and then work out what measurable or easy perceptible attributes can be derived from this[301]. But what are the water management objectives?

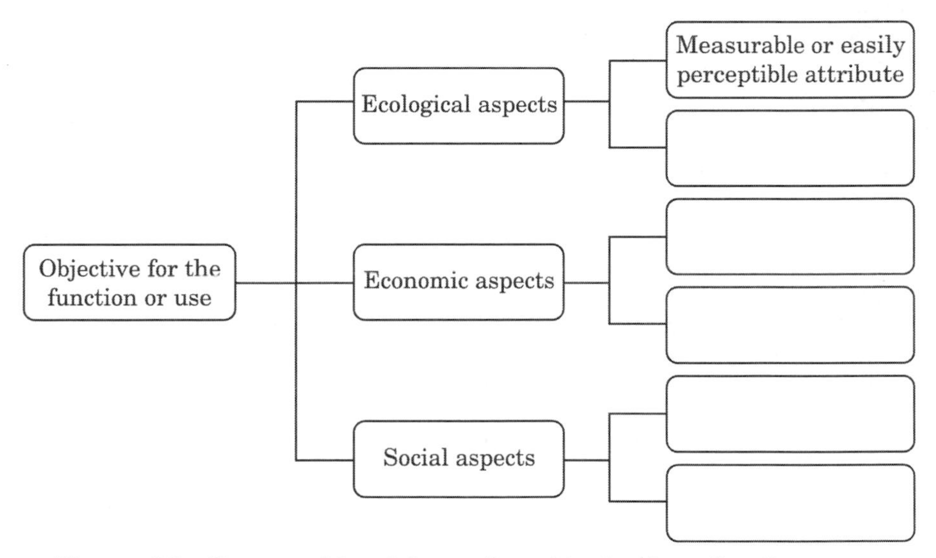

Figure 5.3 Structured breakdown of an objective for a function or use

Current water management is based on the concept of Integrated Water Resources Management (IWRM) that must balance between the ecological, social and economic dimensions[5, 71, 161, 315-317]. This triangle of dimensions will be used as the basis for defining the information needs for the fundamental objective (Fig. 5.3); each of the three elements is an aspect of a water management objective. By listing the individual objectives for the functions and uses and considering for each objective the ecological, economic, and social aspects the objectives can be worked out into their constituting elements that can subsequently be translated into measurable or easily perceptible attributes. The subdivision into aspects is used to limit the amount of information and to generate imagery[231]. Distinction between the three different aspects is central to an integrated approach (also see Box 5.4). The breakdown is not necessarily restricted to three steps, like the ones shown in the figure; it may be useful to distinguish between sub-aspects and build a larger structure that clarifies the breakdown.

Box 5.4 Four capitals framework[4]

The four capitals framework that supplements the three elements of Integrated Water Resources Management (IWRM) with human-made aspects like infrastructure can be applied as an integrated description of the functions and uses of a water body. The four capitals framework emerges from a study published by the World Bank. Four types of 'capital' are distinguished in this study to be able to define sustainability: 1) Natural capital; the stock of environmentally provided assets that provide a flow of useful renewable and non-renewable goods and services, 2) Human-made capital; the 'fabricated' capital like machines, factories, buildings and infrastructure, 3) Human capital; the 'stock' of people with a specific level of education, health and nutrition, and 4) Social capital; the institutional and cultural bases for a society to function. These capitals are interconnected; decrease from one capital (for instance extraction of oil decreases the natural capital) leads to an increase of another capital (for instance oil used to make asphalt roads increases the human-made capital). The total capital in this example could remain the same. Levels of sustainability in this definition are based on the way these capitals are managed. 'Weak' sustainability is maintaining the total capital intact, 'sensible' sustainability adds to this some concern to the composition of the capital, 'strong' sustainability is when different sub-components of the capitals are kept intact, while 'absurdly strong' sustainability would never deplete anything (non-renewable resources cannot be used at all). The four capitals framework could be included into the structured breakdown framework instead of the water management objectives framework.

In a study conducted to develop a monitoring network within a program to restore brackish gradients along the coastal area of The Netherlands, an overview was made of the possible socio-economic aspects of gradient restoration next to the aspects that would be monitored to determine the ecological goals of the restoration projects[301] (see Table 5.1). These socio-economic aspects were not further elaborated. The table nevertheless provides an overview of possible aspects and the goals that could be attached to these aspects. For instance, the economic effects for agriculture can be translated into goals like a certain percentage in change in function of arable land or percentage of change of crops towards salt resistant crops. For recreation, goals can be set like increase in recreational possibilities like fishing, bathing, etc.

Table 5.1 Overview of possible aspects of uses to be monitored (selection from full table[318])

Aspects uses	*Water management*	*Use*	*Opportunities*	*Threats*	*Compensating measures*
Agriculture	Soil becoming brackish, washing through, water discharge	Land use, choice of crops, sprinkle water	Salty crops, management compensation	Change in function, restricted choice of crops, reduction in bird populations	Exchange of land, management compensation, farm relocation
Recreation	Transport routes, constructions, falling clear of water, fouling	Accessibility, peace and quiet, space	Falling clear of water, adventure, variation	Falling clear of water, construction, flow	Construction of new facilities
Fisheries	Fish migration, population structure, water depth	Offshore fishing, fresh water fishing	Migration and population structure, diversity, spawning grounds	Changes and fresh—salt water fishing	Technical innovation
Shipping	Water depth, transport routes, constructions, fouling	Transport, harbors, trans shipment		Navigable depth, transport routes, constructions	
Drinking water abstraction	Chloride content of ground water	Ground water abstraction	Chain of water management	Water becoming brackish	Supply from another source

5.3 STRUCTURED BREAKDOWN OF CAUSE-EFFECT RELATIONSHIPS

The second line in the integrating decision model is to make an inventory of problems and opportunities that influence reaching the objectives for the functions and uses. This defines how the natural environment (for instance the biophysical conditions) and human uses of the water resources limit or enhance reaching the objectives. It is also the basis for the compliance function; standards are set to respond to problems or to secure uses. At this line, the information user ponders: "What do I know about positive and negative factors that promote or hinder the functioning or use of the water system"? A different type of reasoning is required here as compared to the first question, as it now deals with indicating causal relationships. The Driving forces-Pressures-State-Impacts-Responses (DPSIR) indicator framework will be used for the breakdown of this second line. Before going into detail on the breakdown, a short overview on what indicators are, and of the DPSIR framework and how it developed will be given.

5.3.1 What are Indicators?

Indicators are useful in communicating between the different worlds of information users and information producers because they can present information in a condensed/aggregated format and are often linked to specific problems or issues, which are in turn based on specific management needs. Simplification and quantification of information on the water system under concern makes this information accessible and comparable to other water systems[319-322]. Indicators are defined here as an observable or measurable quantity/variable/parameter, representing a process in the environment and having significance beyond its face value[323]. Ideally, through indicators, tacit knowledge is converted into explicit knowledge as the choice of the indicator is done on the basis of expert (tacit) knowledge (Box 5.5). The construction of an indicator is a means of achieving reduction in data volume while retaining significance for particular questions.

Box 5.5 Indicators and indices

Indicators and indices are often confused. An index is a statistical concept (a set of aggregated or weighted determinands or indicators) that provides an indirect way of measuring a given quantity or state. Effectively it is a measure, which allows for comparison over time. The main point of an index is to quantify something that cannot be measured directly, and to measure changes. An index can be used as an indicator (like the Dow Jones index) and indicators can be used as part of an index[321, 324].

An indicator can range from the simplest variable to complex, aggregated functions, but it's most important concept is that it presents a quantitative picture that reflects a complex process. Indicators should quantify information so their significance is more readily apparent and they should simplify information about complex phenomena to improve communication. Note that policy makers often prefer simple indices while most scientists prefer more complex sets of variables. As scientists develop indicators, most indicators should represent empirical models of reality, be analytically sound, and have a fixed methodology of measurement[325]. The result of the development of indicators is often that they are at best a complex aggregation of variables and at worse no more or less than the notorious, extensive list of variables.

Nevertheless, indicators may be useful when they are user-driven, policy-relevant, recognizable and clear. User-driven implies that the user specifies what (type of) indicators should be used. There can therefore not be a fixed set of indicators to be used for fixed problems, but for each specific situation, specific indicators are chosen. Policy-relevant implies that indicators should support the policy-making and policy-evaluation processes. This in turn leads to indicators that describe the (entire) context of the problem and include aspects of the (full) socio-economic system. Finally, indicators should be recognizable and clear, implying that the people the indicator is addressing know what message is contained in the indicator[319, 320, 326, 327]. Indicators should enable or promote information exchange regarding the issue they address[328]. And, indicators should give an indication of a specific situation, not describe a process. Indicators will however only serve this purpose if they are based on an analysis of the actual problem and on the information needs derived from this analysis.

Indicator frameworks provide a typology of indicators that are built upon a certain idea and as such form a conceptual model of a problem situation. The framework describes the types of indicators that should be used to describe a certain issue, where each type represents a viewpoint. Individual indicators can be chosen to describe each type in the framework. The Driving forces-Pressures-State-Impacts-Responses (DPSIR) framework for instance describes a problem situation as a cause-effect chain. This framework will be further elaborated in the following section.

5.3.2 The DPSIR Indicator Framework

The DPSIR framework for environmental indicators as developed by the European Environment Agency is a well-known and well-developed framework and will be discussed here to see how an environmental indicator framework relates to the policy-analysis as described by Dunn[130]. The framework has been developed from the Pressure-State-Response (PSR) framework[310, 327] that starts off from the causality chain and finds its roots in the ecological subsystem of the socio-economic system. The human subsystem (or the socio-economic system) puts *pressures* on the environmental subsystem through pollution and resource depletion. The result is a specific *status* of the environmental subsystem. As a result of this status, the environmental subsystem replies with natural feedbacks, while there is also a societal *response* from the human subsystem. To give an example; fisheries subtract resources from the environmental subsystem to feed the human subsystem. As a result, the fish stock will change, also having an impact on the ecosystem as a whole because of the influence on for example birds that predate on fish. The societal response in this case can be to impose fish quota. Indicators can be chosen to measure the pressure, status and responses.

Van Harten and others[329] later supplemented this framework by adding the Impact on the environmental functioning of the status of the environment, like a loss of biodiversity or reduced purification capacity, into a PSIR-framework. This classification is based on an ecological stress-response chain[330]. A *pressure* indicator describes the intensity of human activities that cause changes in quality and/or quantity of the water system. The *state* indicator describes the status of the quality and/or quantity of the water system. A pressure may result in a new state. The *impact* indicator describes the influence of the state of the water system on functions and uses of the water system. When functions or uses are affected, a societal response may be expected. The *response* indicator describes this. The response aims at a new balance of the water system and the protection of functions and uses. To give an example; suppose the issue is the influence of liquid manure on a groundwater system in relation to the use of ground water as a source of drinking water. An indicator for the pressure on the ground water system could be the nitrogen load of the soil (kg/ha/yr). The NO3- concentration in shallow ground water can be selected as an indicator for the state of the system. The impact indicator may be the suitability for use of the ground water as a source of drinking water. A response indicator could be the percentage of the total amount of liquid manure produced that is processed in a way that is not harmful to the environment. If the issue is fully recognized, policy makers will focus mostly on status and response indicators.

Finally the EEA added the driving force, giving a closer link to the human activity. According to the view represented in the DPSIR-framework, social and economic developments are the Drivers (or Driving forces) that put Pressure on the environment resulting in a change of the environmental State. This changing State imposes Impacts on human health and ecosystems. Such unwanted Impacts induce societal Responses that feed back to the Driving forces, Pressures, State, or Impacts depending on the action taken (Fig. 5.4)[206, 328, 331, 332].

Examples of Driving forces are the sources of a problem, like households or agriculture, and the way they produce or use the problem, like waste water or application of fertilizers. The resulting Pressures may be the waste water that flows into the surface water, treated or untreated, or the amount of nutrients that runs off from the land into the surface water or seeps into the ground water. The State of the surface or ground water is then depicted by concentrations of nutrients or organic matter. The Impact can be seen in changes of biodiversity or changes in use, like a drinking water company that has to apply extra

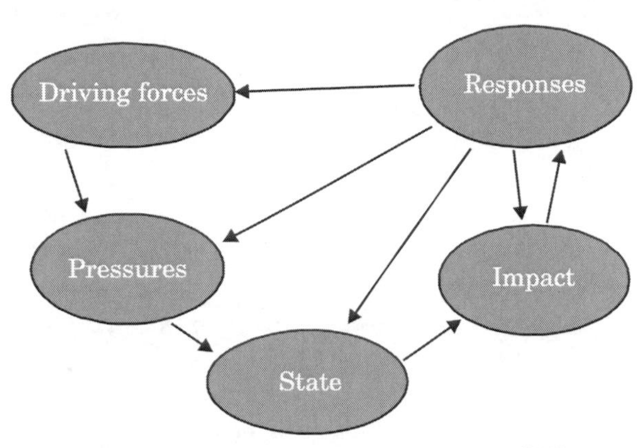

Figure 5.4 The DPSIR framework[331]

treatment. The societal Response finally is the policies and measures taken. Aiming at Driving forces this may be done by, for instance, promotion of best agricultural practices. Aiming at Pressures, applying improved waste water treatment may be a good measure. An example of a measure to change the Status is application of calcium to fight acidification of lakes. Remediation of Impacts is done by, for instance, active fishing of specific fish species that maintain turbidity of the water by stirring up sediments[69].

5.3.3 The Second Line in the Integrating Decision Model

Indicator frameworks are conceptual models that offer support in the structured breakdown of objectives by providing a comprehensive set of aspects. The DPSIR framework is applied here as the conceptual model for the cause-effect relationship (Fig. 5.5). The aspects of the factors of influence are now formed by the five elements of the framework. In addition to the aspects that have been listed to specify the objectives that are related to the uses and functions, and the general description of the situation in the water body as described above, the Driving force gives an exact description of a specific human use or activity. The Pressures describe the influences this use or activity has on the water system. The State describes the existing situation of the water body as related to the desired situation and therefore links back to the attributes as defined in the first line of objectives. The Impact describes how the Pressures and Status influence reaching the objectives. The Response finally describes how the Impacts are counteracted or the Driving force, Pressure and State are influenced. Similar to the aspects connected to each objective, for each problem or opportunity for the aspects Driving force, Pressure, State, Impact, and Response measurable or easily perceptible attributes are identified[78]. Table 5.2 provides an overview of a selection of indicators based on the DPSIR indicator framework. The indicators listed in the table are derived from a range of different indicators as specified in different projects.

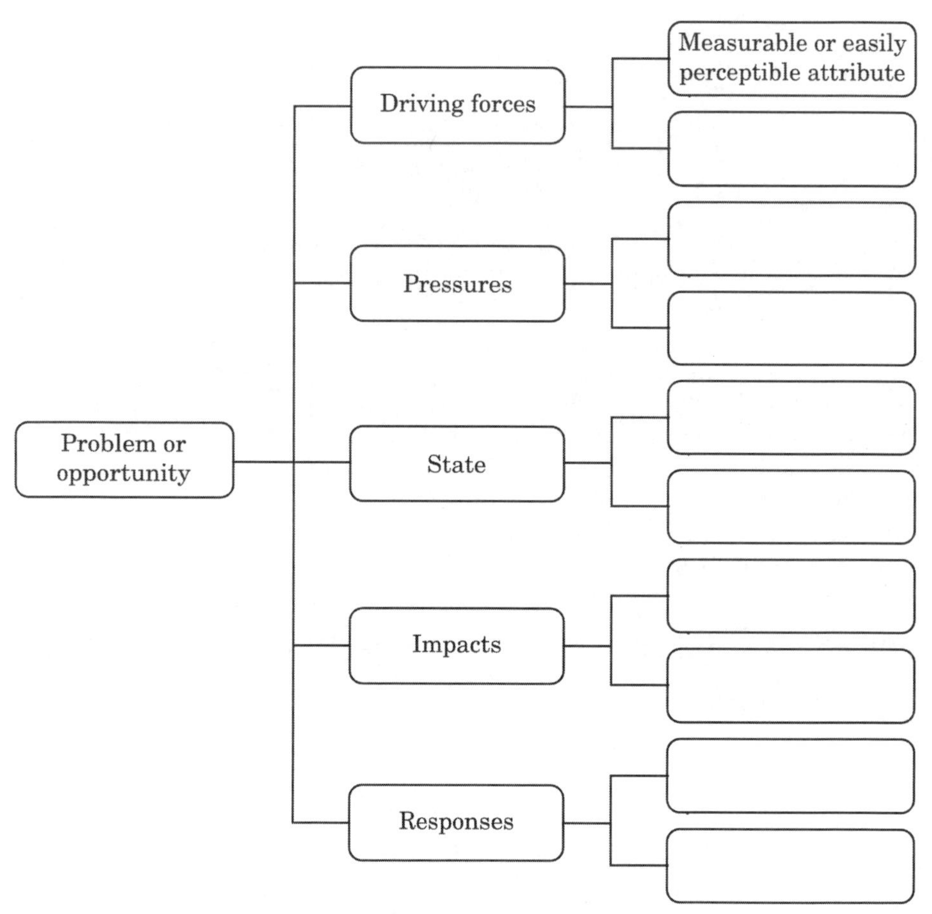

Figure 5.5 Structured breakdown of cause-effect relationship based on the DPSIR indicator framework

Table 5.2 Example of a selection of indicators for several functions based on the DPSIR indicator framework (This table is based on the results of the following reports:[333-339])

	Driving force	*Pressure*	*State*	*Impact*	*Response*
Drinking water[1]					
Organic pollution[2]	Number of (unconnected) households	Load of (untreated) domestic waste water	Oxygen and organic substances concentrations	Costs for removal of organic substances and re-aeration	Investments for connection to WWTP's
	Industries producing organic loads (for instance sugar mill)[3]	Total organic load from industries	Oxygen and organic substances concentrations	Costs for removal of organic substances and re-aeration	Investments in treatment of industrial waste water
Eutrophication	Use of fertilizers in agriculture	Load of N and P washed off into the river	N and P concentrations		Investments into good agriculture practices
	Number of households	Load of N and P from WWTP's	N and P concentrations	Costs for removal of N and P	Investments in treatment capacity of WWTP's
Dangerous substances pollution	Industry that emits dangerous substances	Loads of dangerous substances	Concentrations of dangerous substances	Period of time that water is not usable for production of drinking water[4] The drinking water contains substances that are dangerous for human health	Emissions permitted
	Dump sites	Load from leakage of dump site	Concentrations of dangerous substances	Period of time that water is not usable for production of drinking water	Investments into cleaning and isolation of dump sites

Table 5.2 (*Contd...*)

Table 5.2 (*Contd...*) Example of a selection of indicators for several functions based on the DPSIR indicator framework (This table is based on the results of the following reports:[333-339])

	Driving force	*Pressure*	*State*	*Impact*	*Response*
Accidental pollution	Dump sites - flooding	Load from flooding events	Concentrations of dangerous substances	Period of time that water is not usable for production of drinking water	Investments into cleaning and isolation of dump sites
	Industrial accidents	Loads from accidents	Concentrations of dangerous substances	Period of time that water is not usable for production of drinking water	Investments into prevention of accidents
Industrial uses					
Organic pollution	See drinking water	See drinking water	See drinking water[5]	Period of time that water is not usable for industrial use	See drinking water
Eutrophication	See drinking water	See drinking water	See drinking water	Period of time that water is not usable for industrial use	See drinking water
Dangerous substances pollution	See drinking water	See drinking water	See drinking water	Period of time that water is not usable for industrial use	See drinking water
Accidental pollution	See drinking water	See drinking water	See drinking water	Period of time that water is not usable for industrial use	See drinking water
Irrigation					
Dangerous substances pollution	See drinking water	See drinking water	See drinking water	Period of time that water is not usable for irrigation	See drinking water

	See drinking water	See drinking water	See drinking water	Period of time that water is not usable for irrigation	See drinking water
Fishing					
Organic pollution	See drinking water	Load of (untreated) domestic waste water	Oxygen conditions	Fish kills non-hatching of eggs	Investments for connection to WWTP's
		Bacterial loads	Bacteria concentrations	Fish not consumable	Investments for WWTP's
	See drinking water	Total organic load from industries	Oxygen conditions	Fish kills non-hatching of eggs	Investments in treatment of industrial waste water
Eutrophication	Use of fertilizers in agriculture	Load of N and P washed off into the river	N and P concentrations	Turbidity changes in fish populations	Investments into good agriculture practices
	Number of households	Load of N and P from WWTP's	N and P concentrations	Turbidity changes in fish populations	Investments in treatment capacity of WWTP's
Dangerous substances pollution	Industry that emits dangerous substances	Loads of dangerous substances	Concentrations of dangerous substances, both water and sediment	Fish diseases Fish not consumable	Emissions permitted
	Dump sites	Load from leakage of dump site	Concentrations of dangerous substances, both water and sediment	Fish diseases Fish not consumable	Investments into cleaning and isolation of dump sites

Table 5.2 (*Contd...*)

Table 5.2 (*Contd...*) Example of a selection of indicators for several functions based on the DPSIR indicator framework (This table is based on the results of the following reports:[333-339])

	Driving force	*Pressure*	*State*	*Impact*	*Response*
Accidental pollution	Dump sites - flooding	Load from flooding events	Concentrations of dangerous substances	Fish kills Fish diseases Fish not consumable	Investments into cleaning and isolation of dump sites
	Industrial accidents	Loads from accidents	Concentrations of dangerous substances	Fish kills Fish diseases Fish not consumable	Investments into prevention of accidents
River regulation	Tons transported by ships	Changes in river flow regime	Ecotopes	Changes in fish population	investments into river restoration
Recreation					
Organic pollution	See drinking water	See drinking water	Aesthetic properties of water (odor, color)	Period of time that no recreation is possible	See drinking water
		Bacterial loads	Bacteria concentrations	Number of people becoming ill	See fishing
Eutrophication	See drinking water	See drinking water	Aesthetic properties of water (odor, color)	Period of time that no recreation is possible	See drinking water
Dangerous substances pollution	See drinking water	See drinking water	Concentrations of dangerous substances, both water and sediment	Number of people becoming ill	See drinking water
Ecosystem functioning					
Organic pollution					
Eutrophication					

Dangerous substances pollution	Etc. ....
Accidental pollution	
River regulation	
Land drainage	
River regulation	Etc. ....
Impact on receiving water body	
Organic pollution	
Eutrophication	
Dangerous substances pollution	Etc. ....
Accidental pollution	

[1] For each of the water management targets, an item is inserted into this table

[2] For each of the water management problems connected to the target, an item is inserted into this table

[3] Any water management problem can have different sources (driving forces)

[4] Indicators chosen should be related to the description of the uses, problems and measures

[5] Reference to other parts of the table should only be made when there is no difference between the parts. If, for instance, special water quality standards for industrial uses are set, data analysis will be different. Such differences have to be accounted for when specifying information needs. The indicator can be the same, but then parameters may differ.

Box 5.6 Ecological model used as alternative for the DPSIR model

In the study to develop a monitoring network within a program to restore brackish gradients as described in Section 5.2 an ecological model was used to cover for the structured breakdown of cause-effect relationship[301]. The scheme as shown in Fig. 5.6 was used.

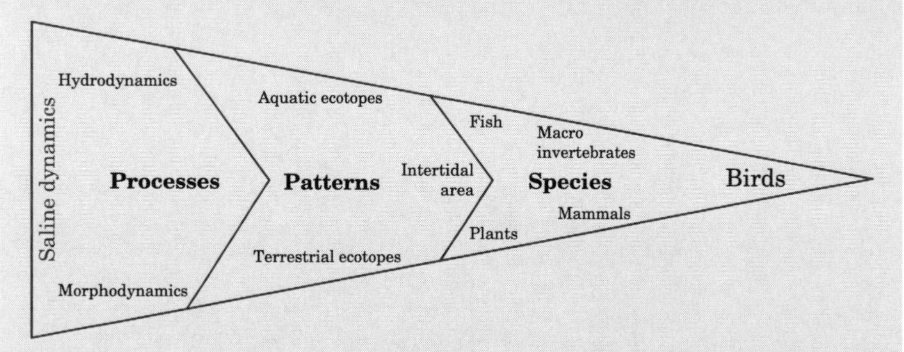

Figure 5.6 Hierarchic connections between hydrological and morphological processes, patterns in ecotopes and occurrence of species.

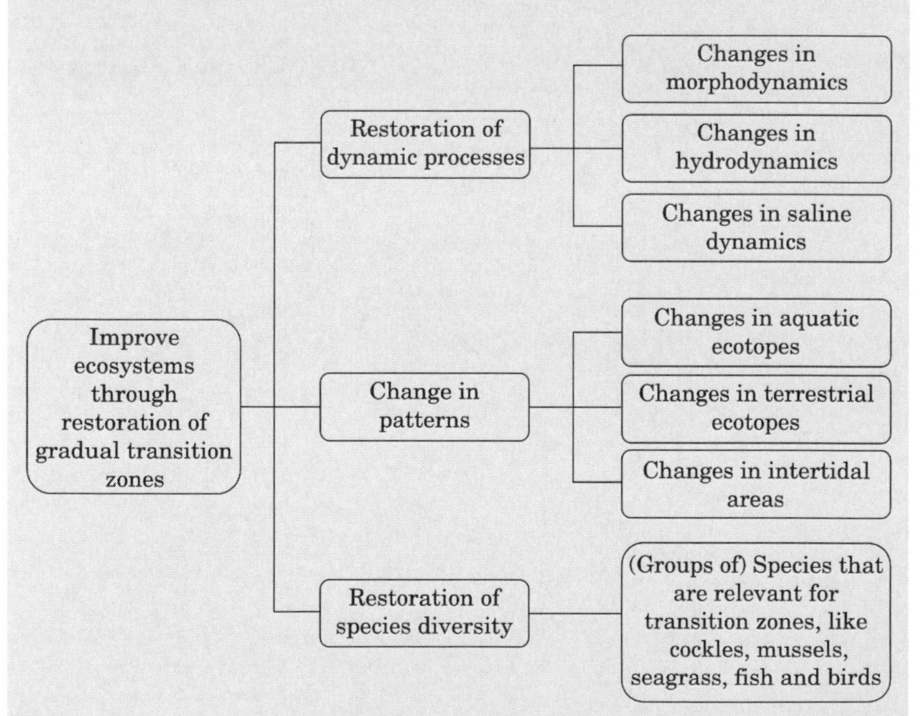

Figure 5.7 Systematic breakdown of cause-effect relationship based on the hierarchic connections between hydrological and morphological processes

This scheme as used by ecologists represents the hierarchic connection between the various aspects discerned in the information about the ecosystem. Largely the figure depicts that without the process, the patterns will not develop and without the patterns developing, the species will not be present. Use of this model proved very helpful in further defining the objectives of the development of gradients. Systematic breakdown of this scheme into the proposed improved method[301] leads to the next scheme where the fundamental objective of restoration of ecosystems as the rectangle at the left of the figure can be divided into three aspects of ecosystems described in rectangles in the middle that each in turn can de described by the measurable criteria listed in the right rectangles (Fig. 5.7):

The major advantage of using such a familiar framework is that it links to the practice of the information users. The major disadvantage of using this model was that it only showed the cause-effect relationships between the ecological components. The social and economic aspects were not included in the process and as a result had to be included afterwards.

5.4 STRUCTURED BREAKDOWN OF MEASURES

The third line in the integrating decision model is to identify measures directly related to problems or more generally related to the functions and uses. This gives insight into what is already done and what can and should be done in future. The question here is: 'What do I know about the implementation of measures and their effectiveness?' This requires reasoning in terms of relationships between objectives and means (Box 5.7). The information user must be able to make an analysis of reasons why the objective is not achieved and the actions that can be taken to remedy the problems. The Responses, as listed in the second line in the integrating decision model, are related to the aspects of the fundamental objectives to be reached, together with measures that are not linked to problems. This is largely a bottom-up approach. The two aspects to be addressed are 1) if measures are implemented and 2) if the measures have been effective in reaching the objectives (Fig. 5.8). The auditing function of monitoring is addressed by identifying if measures are actually taken; the accounting function is addressed by looking at the effectiveness of the measures. Often, measures are implemented only partly or even not implemented at all. There may be valid reasons for not implementing measures. Nevertheless, for the policy maker it has to be clear if non-implementation is the major reason that the problem situation does not change or that there are other reasons why the situation is not changing. When the measure is implemented, it may be possible from the changes as monitored through the aspects as mentioned in the previous sections to assess the effectiveness of the measure. Similar to the breakdown structures described above, for each measure for the aspects implementation of the measure and effectiveness of the measure attributes are identified (Box 5.8).

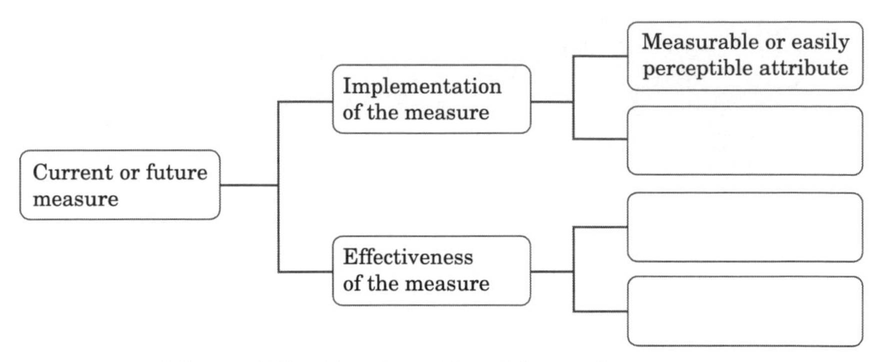

Figure 5.8 Structured breakdown of measures

Table 5.3 provides an overview of the measures that were anticipated for the policy on restoration of gradients and the expected ecological effects from these measures[301]. In this table the structured breakdown of measures can be applied. For example, to assess the implementation of the measures, the number of fish ladders or the number of estuaries that are restored can be monitored. Then, to assess the effectiveness, monitoring of the fish migration shows the extent to which the measure has had an effect. The monitoring of measures can be effectuated in this way.

Table 5.3 Overview of measures and expected effects[318]

Measures	*Expected ecological effects*
Adapted operation of sluices and adapted drainage regime	Increased fish migration More stable brackish water zones Reduction in fish skin ulcers
Construction of fish ladders	Increased fish migration
Adapted management of marshes	More stable marshes with all different successional stages Protection of dikes Expected vegetation and gradients in vegetation
Development of structures	Sea grass beds Mussel banks, other shellfish banks Species-rich green embankment slopes Natural banks
Restoration of estuaries and estuarine transitions	Dynamic—stable fresh water—salt water gradient Unhindered fish migration Brackish water flora and fauna Absence of fish diseases
Restoration of brackish areas within floodplains	Brackish water flora and fauna

The three frameworks are overlapping to some extent; the three elements of the IWRM framework, especially the ecology element, are closely linked to the State information in the DPSIR framework, while the Response information is further defined in the structured breakdown of measures. As the perspective in each framework is different it will however add to a comprehensive set of information needs. Any overlap will become clear from rearranging the information needs for the development of an information strategy as the next phase in the information cycle[340]. The integrating decision model complemented with the three frameworks thus gives insight into the effectiveness of the policy decisions and indicates if additional measures and or revised policies are needed.

Box 5.7 Overweighing of objectives

Bana e Costa and Beinat[311] caution that often several attributes, which are in fact alternative descriptors for the same key-concern (or fundamental objective), are included in the problem analysis. This leads to overweighing of that key-concern in the evaluation process. This danger also exists in water quality monitoring where large numbers of individual determinands are measured that each influence the ecological quality of a water body.

Duplication of determinands can occur from combining different objectives. Through the structured breakdown scheme it is obvious that if an objective is no longer valid, the determinands attached become obsolete. Nevertheless, the same determinands may still be relevant through another objective, and the schemes in this way provide a structured way of reporting that can easily be changed while the consequences of changes can be traced. Similarly, going upward in the hierarchy and choosing another viewpoint can lead to choosing another set of attributes, without changing the initial concern. A parameter-oriented monitoring network can in this way be altered into for instance a more effect-oriented network. This is discussed in more detail in the following chapter.

Box 5.8 Requirements for MCDA

Belton and Stewart [294] distinguish between value relevance, understandability, measurability, redundancy, judgemental independence, operationality, completeness and conciseness, and simplicity versus complexity as identifying attributes for the different criteria to evaluate. These considerations are drawn up for the purpose of Multi Criteria Decision Analysis (MCDA) and not all are essential when specifying information needs. Redundancy in information needs for instance will be easily identified when developing a monitoring network. If the same parameters are needed for different purposes, they put different requirements on for instance the measuring frequencies or locations. By comparing these different requirements, an optimum is found for that specific attribute, while the use of the resulting data serves different purposes. It is nevertheless important to critically review the information needs on the basis of the abovementioned considerations in view of effective and cost-efficient information production.

5.5 WORKING SCHEME

The working scheme summarizes the overview of information needs and forms the basis for the information network. Based on the integrating decision model as described in Section 5.1, for each objective a working scheme is produced in which the information needs are reported. Table 5.4 provides a format of such a working scheme. The information is about objectives, positive and negative factors of influence, and the implementation and effectiveness of measures. These are the attributes that should provide the integrated information on the fundamental objectives.

Table 5.4 Working scheme for information needs (intentionally left blank)

Matrix of information needs per objective			
	Required determinands	***Required processing***	***Further details***
Attributes for the objectives			
Attributes for the positive/negative factors of influence			
Attributes for the implementation and effectiveness of measures			

When the attributes connected to each objective are known, they have to be further specified. A first step is to determine the required determinands to quantify the attribute. For standard testing of chemical substances for instance, not only the limit value but also requirements for the number of data to be used and the calculation method are prescribed. Also, for comparability reasons, a transformation to standard conditions may be necessary. For instance, to make concentrations of for instance metals and organic substances comparable over different locations, contents of organic carbon, lutum, and suspended solids of the water at these locations are needed as these influence the measured concentrations. Calculations are needed to correct for the composition and to transform the analytical data to a standard. The required processing thus has to be specified. This also concerns the use of mean or median values, or in case of loads the calculations method, or when predictive models are needed, the type of model to be used. Then there are several other details to be specified, like the relevant margin, indication of locations and frequency, the type of presentation to be used, etc. Such details are included in the working scheme.

Table 5.5 provides examples of how the working scheme can look like after working through the structured breakdown on the basis of the integrating decision model. Note that still not all elements of the integrating decision model are included in this table.

Table 5.5 Working scheme for the aspect Ecology of a project for the monitoring of the River Meuse[341]

Water management target[6]	***Information question***	***Monitoring target***	***Parameter***	***Accuracy***	***Method/ technique***	***Operational characteristics***		***Connected to***	
						Location	***Time scale***	***Other themes***	***Other organizations***
Conservation of size and quality of existing natural values Development of ecotopes (connected to the river) together with landscape determining factors	What is the share and spatial distribution of natural areas in the winter bed of the river Meuse? What is the status of realization of nature development projects (surface area and location)?	Policy evaluation Legal obligation	Surface area and location of areas with natural management Surface area and location of purchased and devised reservation- and nature development area	Ha ± 5% Scale 1:100.000	GIS	Covering the area	Annual	Safety Landscape	Province
	What are the existing river ecotopes and does this comply with the targets? What is the biological	Policy evaluation Development of knowledge	Surface area and location of river ecotopes Biological quality of ecotopes, based on the	Ecotopes: 1:10.000 Signaling of trends on the level of	Aerial photo mapping/ ecotopes Field inventory species	Covering the area Representative locations in summer bed and	Once per 4 years Depending on species group monthly to annual to	Water quantity Morphology Landscape	Province

Table 5.5 *(Contd...)*

Table 5.5 (*Contd...*) Working scheme for the aspect Ecology of a project for the monitoring of the River Meuse[341]

Water management target[6]	***Information question***	***Monitoring target***	***Parameter***	***Accuracy***	***Method/ technique***	***Operational characteristics***		***Connected to***	
						Location	***Time scale***	***Other themes***	***Other organizations***
	quality of existing and new developed nature- and management areas? Which changes occur in presence of protected and target species		following species groups: Minimal: plants, macro-invertebrates, fish, waterfowl; Optimal: amphibians, butterflies, mammals Presence of protected and target species	water system	groups/ protected species Analysis of biological quality	winter bed	once every 4 years		
Development of an ecological connection zone river Meuse	What is the status of acquisition and upgrading of nature friendly riverbanks and spawning grounds? Do these banks	Policy evaluation Legal obligation	Length and surface area of acquired and upgraded nature friendly river banks Biologic	Ecotopes: 1:10.000 Signaling of trends on the level of water system	Atlas/ ecotopes-GIS Field inventory and analysis of biologic quality	Covering the area Representative locations	Once per 4 years Depending on species group monthly to once every 4 years	Water quantity Morphology Landscape	Province

Table 5.5 (*Contd...*)

	and spawning grounds function?		quality, especially of these species groups: Water and bank plants Waterfowl Macro-invertebrates Fish						
Emphasize the water system and river basin management approach	Are the ecological networks functioning? Are sustainable populations of target species possible?	Policy evaluation Develop-ment of knowledge	Quality ecologic network Guiding species	Signaling of trends on the level of water system	Analysis of network relations	River basin	Once per 4 years	Landscape	Province
Dutch part of the river Meuse is fully accessible for fish	Are fish passages realized and do they function?	Policy evaluation Risks for manage-ment	Actual fish passages Movements of fish		Fish sampling, transmitter	Covering the area	Starting situation, first years after realization	Water quality	Province Water board
Canals: Realization of ecologic connection zone along the river length	How far have realizations of nature friendly riverbanks come and do they function? Are migration bottlenecks	Policy evaluation Develop-ment of knowledge	Length and surface area of acquired and upgraded nature friendly river banks	Scale 1:10.000 Number and phase of realization	Aerial photo mapping/ ecotopes Field inventory species	Some represen-tative locations	Starting situation Once per 4 years	Water quality	

Table 5.5 (*Contd...*) Working scheme for the aspect Ecology of a project for the monitoring of the River Meuse[341]

Water management target[6]	*Information question*	*Monitoring target*	*Parameter*	*Accuracy*	*Method/ technique*	*Operational characteristics*		*Connected to*	
						Location	*Time scale*	*Other themes*	*Other organizations*
Removal of barriers in cross-direction	inventoried and solved?		Biologic quality based these species groups: Water and bank plants Waterfowl Macro-invertebrates Fish Situation of resolving migration bottlenecks		groups Probe testing Species lists Counting				
Dynamic river processes are undisturbed and natural	Are hydromorpho-dynamics occurring? What is the minimum and mean discharge? What is the stream velocity and duration of flooding?	Policy evaluation Risks for manage-ment	Erosion Discharge Stream velocity Duration and frequency of flooding	Information should provide insight into periodic changes	See themes morphology and water quantity				

Water quality causes no limits for ecological functioning of the water system	Does the water quality comply with ecologic standards (see theme water quality)?	Policy evaluation	See theme water quality	See theme water quality	See theme water quality
Water (sediment) quality causes no risks for functioning of ecosystems	Does the water sediment quality comply with standards? What are the eco-toxicological risks?	Policy evaluation Develop-ment of knowledge	See theme water sediment quality	See theme water sediment quality	See theme water sediment quality

[6] Management targets like the ones mentioned in Table 4.2 are listed here

5.6 EXERCISES

- Based on the water management analysis as performed through the exercises in Chapter 4, determine what the integrating decision model looks like. Make the distinction within the policy objectives between goals, problems and measures. For each of these three elements, determine the aspects that should be taken into account.
- Determine the policy goals and make a division of these goals over the ecological, economic and social aspects. Parts of policy goals can be attributed to one or more of the three aspects. After making this division, identify measurable attributes for each of the three aspects. Develop an information needs hierarchy, based on the results.
- From the function/issue table developed in the previous chapter, identify the problems attached to the policy objectives. For each problem, elaborate on the DPSIR indicator framework; determine the Drivers, Pressures, Status, Impacts and Responses for each problem. Develop an overview table for each problem and distinguish between the different types of indicators.
- For all the policy measures as identified, determine what attributes can be attached to them to identify if they have been implemented. Also determine the expected effects of the measures and the way these effects can be assessed. Attach measurable attributes to the effects of the measures.
- Looking at the overview of aspects you have developed in answering the previous questions, if you would have the information on all these aspects available, would you be able to take the decisions you want to take?

6

The Next Steps

The results of the entire rugby ball process have now been documented in the form of a comprehensive overview, a blueprint for the new information network, or a program of action, as described in the previous chapter. This result will target decision makers rather than monitoring people as it forms the translation from the water management objectives into information needs. It is also the preparation for the next steps, which need approval from but have less involvement of the decision makers. The following step in the information cycle is to transform the results into a form that is usable for the development of an information network, and that therefore targets the monitoring people. The transformation has consequences for the link between the water management objectives and the information produced; the focus will be on the information production issues while the objectives run the danger to become detached.

This chapter deals with a general description of finalizing the first step and turning to the next step in the information cycle. The focus will be on how to maintain the link between the information production and the water management objectives without going into much detail. An overview will be given of the issues that have to be dealt with in defining an information strategy. Section 6.5 provides an example of an information strategy as developed in a study performed in The Netherlands in the Meuse River. After studying this chapter the reader will have obtained the following abilities:

- Understand the necessity to document the results of the process
- Understand the role of the results in the overall process
- Be able to transform the information needs into a basis for an information strategy

6.1 DOCUMENTING THE RESULTS

As discussed in Chapter 3, the results of the entire rugby ball process and agreements of the Concluding workshop need to be documented. This is necessary to create a solid basis for the design of an information collection network. Designing the information network is the obvious reason to specify the information needs. This is how it fits in with the concept of the information cycle; the information needs as specified are input for the development of the information strategy. Next to that, a well-documented overview of the information needs is needed to be able to assess, after some time, whether the information as specified is actually produced. Moreover, it will enable to determine whether the information produced provides sufficient basis to evaluate if the objectives will be reached.

Other reasons for documenting the results are that the documentation can be sent to all participants in the process—which provides them with the opportunity to assess the results of their efforts and make comments—and the documentation shows that their input has led to a concrete result—which will motivate them if their input is needed another time in a similar exercise. Last, but no least, the documentation is needed to convince decision makers that the process has produced a solid result that can act as a basis for the next steps. As the decision makers' role will be less prominent in this following phase, with fewer possibilities to influence the process, the documentation is intended to make them confident that the information production process is based on a reliable assessment of the information needs. The documentation is therefore important to be able to close the information cycle once the information is produced, while it also serves as the basis for re-entering the cycle to improve the information production. Next to that it is an important means to build confidence among both decision makers and stakeholders.

Another important but less obvious reason to document the results is to enable reconstruction of the objectives for the information as collected. There are two aspects to this point. One aspect of being able to reconstruct the objectives is that the structure that was built to describe the information needs will not be used anymore in the next step of developing the information strategy, as will be described below. In Chapter 5, the structured breakdown was described to show how a water management objective was translated into measurable attributes. One (or more) of the attributes that were derived in this exercise may also provide information about (one or more) other objectives. During the further development of the information network and the collection of information, the original objectives for collecting the information are usually not included. Only when the information is utilized, for instance, when writing a report, the objectives come into sight again. Documentation of the link between objectives and attributes is important to accommodate for this latter purpose. Moreover, at a certain point in time, an objective may become obsolete, for instance because of a change of policy. The attributes connected to an objective that has been skipped, can be deleted from the information network, unless they have an information value for other objectives. Also in such a situation, documentation of the connection between objectives and attributes is needed.

The other aspect of being able to reconstruct the objectives is that information collection is often under pressure to reduce its costs as discussed in Chapter 1. Regularly, the information network is questioned, the questions generally coming down to 'is all this information really needed?' When the process and results of the information needs specification is properly done, the answer to such questions can be easily answered; the starting points for the information collection are clear, the water management objectives covered are specified, the process that made the transformation from objectives to information needs is clarified, and the people that were involved and that approved of the information needs are known. The available documentation can in such a situation help to direct any discussion about budgets towards the process. If the choices are challenged it can be clarified that the choices were communicated and approved by the participants, where it also clear who were the participants and what their organizational role was. The discussion in this situation is not so much about the validity of the information needs but can be turned to the choices made and if maybe other choices are necessary given the new or changed situation. If there is a need to reduce the means available for monitoring, the discussion can focus on prioritizing certain objectives for which information is vital. This may give room to skip collecting information that is not related to these priority objectives. However, most of the efforts in monitoring networks go into fulfilling legal objectives and there is for that reason usually little room to substantially decrease the budgets for the information network[58]. It may nevertheless be clear that a comprehensive report on the results of the process of assessing the information needs is helpful to avoid vague accusations of collection too much data while not producing the required information; the water information gap, as described in Chapter 1.

6.2 FINALIZING THE RUGBY BALL PROCESS

The blueprint, comprehensive overview, or program of action for the information strategy as developed forms the finalization of the project. It describes how the project was performed and what the outcomes are. The final step is now to discuss the final document with the commissioners. They have to agree that the project was done in the proper way and that the results can act as the basis for the next steps. This section describes the contents of the documentation.

The documentation typically includes a description of the process. This entails the actors that were involved together with the role they played, like the people (and their organizational position) that performed the process and a listing of workshop participants (and their organizational position). Also the commissioners of the process are described. The structure of the process is described, including the timing of the various milestones like start of the project, sets of interviews performed, workshops conducted, etc. The steps in the process are described and reference is made to the documents that were produced during the process like the project plan, interview reports, input documents for the workshops and workshop results. Next to that, the formal and informal decisions are documented. This includes, among others, the approval of the project plan and the way possible feedback to the workshop results was organized. In the latter case, for example, if participants are not responding to the request for comments on the workshop report, this is often considered as approval of the result (consent). Also the content of the process is described. This includes the policy objectives that were taken into account, the water(s) under scrutiny, possible exclusions—issues that were excluded from the process, the intended audience (information users), and other relevant aspects of the starting points. Next to that, the documentation includes the full overview of the breakdown of the policy objectives into measurable attributes as described in the previous chapter.

Ideally, the document of the process is accompanied by a plan for the next step. This plan contains the activities to be undertaken to come to an information strategy, the people that will be involved in the process as well as their role, the anticipated timeline, and the products that are anticipated. After approval from the commissioners, the report as well as the project plan is sent to the people that were involved in the first step and to the people that will be involved in the next step. In this way there is clarity about the outcomes of the process and the way the results will be used for the next steps. An example of the table of content of a final report is shown in Box 6.1.

Box 6.1 Example of an annotated table of contents of the final report

Table of contents:

1. Introduction

The introduction describes the reasons why the project was performed as well as the part of water management that is included in the study. The organizational set-up is described. The introduction also describes other relevant considerations.

2. Delineation of the project

This chapter describes in detail what part of the water management is taken into consideration. It describes the policy objectives that were taken into account, the water(s) under scrutiny, possible exclusions; issues that were excluded from the process, the intended audience (information users) and other relevant aspects of the starting points. Largely, this chapter summarizes the results of the Exploration phase of the rugby ball methodology.

3. Method

This chapter describes how the process was performed, largely describing the steps in the rugby ball and how they were performed. This, for instance, includes if and how the Initiation and Conclusion phase were performed in the form of workshops and how these workshops linked to the other phases of the rugby ball process.

3.1 ***Process description***

The structure of the process is described here, including the timing of the various milestones like start, sets of interviews, workshops, etc. the steps in the process are described as well as reference to the documents that were produced during the process like the project plan, input documents for the workshops and workshop results. It also includes the formal and informal decisions that are documented, like the approval of the project plan, and the way possible feedback to the workshop results was organized.

3.2 *Actors*

All the actors (the commissioners, the workshop participants, the people that perform the process, etc.) involved are described here, including their position and their role in the process. The CATWOE analysis forms an important basis for this chapter.

4. Results

This chapter provides a full overview of the breakdown of the policy objectives into measurable attributes (as described in Chapter 5 of this book). Larger parts of this chapter will consist of tables. Also the filled-in working scheme is represented here.

5. The next step

This chapter contains the activities to be undertaken to come to an information strategy, the people that will be involved in the process as well as their role, the anticipated timeline, and the products that are anticipated.

6. References

A listing of all relevant documents is presented here. This includes all the reports as produced during the process.

6.3 HOW TO TRANSFORM INFORMATION NEEDS TOWARDS AN INFORMATION STRATEGY?

The result from the information needs specification is an overview of the water management objectives and the measurable attributes connected to them. It also includes qualitative requirements for each of these attributes. This is the basis for the next step in the information cycle; developing an information strategy.

The first action in developing an information strategy is to list the attributes and their requirements, disconnecting the attributes from the water management objectives. Note that the same attribute can be linked to several objectives although sometimes with different requirements. The list of attributes should therefore be cleaned to omit duplication of attributes. The different requirements that may be attached to the attribute are assessed and, if possible, the strictest requirements are linked to the attribute. This is because the information resulting from the strictest requirements can usually also be used as information for those objectives that put less strict requirements on the attributes. Requirements are typically connected to the density in space and time of the necessary data collection. If the requirements of different objectives are contradictory, it may be necessary to use two or more sets of requirements for the same attribute. Information on the same attribute should then be collected in two different ways.

As the requirements for the attributes, as coming from the information needs assessment, are largely qualitative, they need to be quantified. This is done, if needed, in consultation with the information users. The basis for the information strategy now has become a list of attributes, along with the requirements attached to the attribute. Note that the link to the water management objectives gets lost in this step.

As described in Chapter 2, various information sources exist, like literature, existing databases and models next to direct data collection (monitoring). Within monitoring, also different approaches are possible, for instance remote sensing, regular sampling, continuous sampling, use of sensors, etc. Some general strategies in water quality monitoring are[127, 264, 274, 342]:

- Variable-oriented: Collecting data of individual variables. This strategy is useful when the number of potentially harmful substances is limited or for examination of specific processes. This is the approach that is generally applied in water quality monitoring.
- Effect-oriented: Data collection is focused on indicative variables that point out if something is going wrong. If there is an indication of some kind, specialized investigation through more targeted monitoring may point out exactly what is going on[343]. Ecotoxicological testing, for instance, may reveal effects that highly exceed the toxicity accounted for by chemical analysis of individual variables[344]. If an effect has been identified, further research may be needed to identify the exact source of the effect. This strategy is useful when there are no dominating problems that may be related to a limited set of variables.
- Source-oriented: Collecting data of sources that cause adverse effects. This strategy may be used for, for instance, licensing

(effluent monitoring) or policy preparation but is also indicative for pressure on a water system. This is also called Pressure-oriented monitoring.

- Achievement-oriented: Collecting data on a policy goal that is set to be achieved within a given time period. The strategy aims at pointing out the distance to the goal or the effectiveness of measures.
- Predictive: Data collection on events that are expected to happen. This strategy is used in flood forecasting and for policy-analysis. The use of models is imperative in this strategy.
- Knowledge-building: Data collection to understand specific processes in the water. Usually this strategy is performed on a short-term basis.
- Tiered testing: Step-wise approach in which additional data are collected only when available data provide insufficient information. The approach of the EU Water Framework Directive that distinguishes between three types of monitoring; surveillance, operational, and investigative monitoring, can to a certain extent be considered as a tiered testing approach.
- Investigative: Collection of data of which the relevance is not yet clear. This may be done for instance by storing samples, that have not been analyzed. If some event (for instance a fish-kill) occurs, these samples may be analyzed to trace back the source of the event.

The results from the information needs assessment will include a mixture of the different strategies, as may be clear when comparing this listing with the principles used in the structured breakdown as described in Chapter 5. Part of the choice for a specific strategy is consequently already made. Each of the different information sources has characteristics that make them more or less useful for the situation under study.

In this step, a reconsideration of the approaches is made from an informational point of view. Criteria that can be used to assess the suitability of the sources are costs, accuracy, precision, relevance, coverage, risks of operation, availability, etc. Also acceptance of a method (credibility) can be a criterion. When the results of certain methods will not be accepted by the decision makers and information users, such methods should not be applied.

On the basis of these criteria, data and information sources can be selected and ranked for their suitability to produce the information needed. This also includes possibilities to combine different attributes. For instance remote sensing can provide a spatial coverage of algal blooms and this information can be combined with data on the water temperature. The remote sensing measurements are not very precise. *In situ* sampling can provide information on the type of algae involved and give precise numbers about the water temperature but do not show the spatial coverage. Depending on the precision of the information needed and other characteristics like the need for spatial coverage, a choice can be made between using remote sensing or *in situ* sampling or a combination of the two. The related costs to produce the data are of course relevant.

An overview of the sources of all attributes and combinations of attributes will form the first draft of the information strategy. This needs to be complemented with more precise information on the frequency of collection (for instance, collecting information from the database of the statistics bureau would be done once a year or less, *in situ* sampling for chemicals is usually done several times per year), sampling locations, input for models, etc. An assessment needs to be made whether the collected data can be linked to each other. For instance, data about industrial effluents and chemical concentrations in surface water should be collected for the same water system to be able to make the connection between the two. While elaborating on these issues, it may become clear that the initial choice about, for instance, the source may lead to unwanted outcomes like the inability to connect, but may also lead to higher costs or inadequate spatial coverage. Iterations in the process may therefore be needed to come to an overall information strategy.

Note that the collection of data is not something to be done by a single entity. For instance industries can be obliged to monitor their own effluents, fisheries associations keep track of certain variables in their water and can give information on the fish communities and catches, and nature organizations often collect data on various species. All these data may merely need processing to be useful for the purposes as described in the information needs assessment report. A specific issue in this is to ensure data quality and comparability; different organizations may use different analysis methods that make comparability between data difficult, while some organizations may not be able to guarantee quality standards for the data. Such limitations can be overcome by, for example, training of volunteers. Also, alternative data collection approaches can be developed, like installing data loggers on ferryboats that provide regular data on specific stretches. All this may reduce sampling efforts.

6.4 DEVELOPING THE INFORMATION STRATEGY

After having defined the information need and after having transformed the information needs into a list of attributes on which data need to be collected, a strategy for collecting information has to be specified. This is the step necessary to define what information has to be produced by the monitoring system and to decide on the type of monitoring needed. For pollution monitoring this may include physical, chemical, biological, hydrological, effluent or early warning monitoring; for policy implementation this may include interviews and study of literature; for socio-economic information this may include interviews and collection of data from other institutes. For example, the information need from the point of view of supplying drinking water entails an information strategy that is divided in an operational early warning monitoring system and different approaches to look for potential threats. These include the 'target compound' approach where priority compounds are selected, based on their unwanted properties and on the likelihood of occurrence (for instance data on upstream production and use), as well as on degradability/retainability in the treatment process; the 'shotgun' approach where low level random hits showing up in the screening techniques for organic micro pollutants are stored in a separate database in a standardized format that is periodically searched for recurring individual unknowns; and the 'toxicity' approach where toxic effects are detected, using some sort of effect-sensor'[345].

The information strategy has to contain enough information for the monitoring network designer to make the design. The information strategy should specify what has to be measured whereas the network design specifies how it has to be measured. The information strategy should include reference to the data analysis and reporting, because this can influence the network design requirements. The information strategy should also be documented. Such a document is to be presented to and approved by the decision makers that defined the information needs. The decision makers should conclude from the information strategy report if the monitoring system will cover their information need.

Elements of the information strategy report are[239]:

1. The information needs that will be covered by the information strategy and, equally important, that part of the information needs that will not be covered by the information strategy.
2. The concept of the monitoring system, i.e. the type of monitoring/information collection (see above), the variables to measure or collect and preconditions for the selection of locations (minimum/maximum distance from border, intake point, etc.) and sampling frequencies (in terms of reliability) for each (group of) variables.
3. The concept of the assessment system, i.e. the calculation methods to be used (for instance for calculation of loads or trends) (preferably international standards) and the use of graphical tools, statistical tools and other tools - like indices - to present data.
4. Considerations on the proposed concept, like preconditions, suppositions, the statistical models used, etc., but also descriptions of the area, relevant industries, scenarios used, etc.

5. The organizational aspects, in which the following questions should be answered:
   a. Which organization will be responsible for what aspect of the monitoring system?
   b. Are changes in the organization necessary?
   c. What are the problems that can hinder the execution of the monitoring system?
6. A plan for the design and implementation of the monitoring network; what are the preconditions, the planning of the next steps and a financial planning.
7. An analysis of the risks; what are the distinguished possible problems that can lead to the failure of the monitoring system.

Again, developing the information strategy is a process that is done in close cooperation between several experts. The outcomes need to be documented to enable easy reconsideration of the choices made. This is again to avoid unproductive discussions about the budgets. It also avoids having to do the whole exercise over and over again. The information strategy furthermore needs approval from the decision makers, as it entails choices about budgets. This is another argument to document the process and outcomes of that step.

Based on the information strategy, data collection plans like the monitoring network are drawn. For each source as selected in the information strategy, a detailed plan has to be made how and when the data is collected and further processed. It may be clear that with this step, the link between the attribute and the original water management objective is even more difficult to maintain. The documentation in the previous steps is therefore imperative to be able to reconstruct the choices made.

6.5 EXAMPLE: INTEGRATED MONITORING PLAN ZANDMAAS/MAASROUTE

This section provides an example of a project to monitor a program that was performed in the 1990's in a stretch of the river Meuse in The Netherlands called 'Zandmaas'. The Zandmaas/Maasroute protection project was defined to improve navigation, which is extensive on this part of the river. The overall goals of the Zandmaas/Maasroute protection project are[346]:

- Increase the level of protection against flooding;
- Improvement of the navigation route;
- In connection with this, exploit chances for nature where possible.

The integrated monitoring plan for water policy contains the information needs to formulate and evaluate the regional implementation of the national water policy. The water policy describes the targets and measures of water management. Important questions for monitoring were:

- What is the status of the water and what are the changes occurring?
- Will the policy targets be met within the appointed period?
- Do the measures lead to the desired results?

The information questions based on the project goals were determined. Themes were chosen to organize the progress of structuring the analysis of information needs. The themes chosen were: Safety and morphological stability; Shipping; Nature; Landscape; Water quality and sludge; Ground water; Agriculture; Recreation; Living, working and infrastructure; Soil quality and mineral abstraction; and Nuisance for inhabitants. The information on nuisance and mineral abstraction was needed for licensing and calculation of the revenues respectively. Figure 6.1 shows the overall process that was performed to define the monitoring strategy, where the 'What' part relates to specifying the information needs, as described in Chapter 5, and the 'How' relates to developing the information strategy as described in this chapter. Initially this structure was at times felt like a straitjacket that hindered development of ideas. The participants at first showed some reluctance to have their ideas pushed into 'pigeonholes', but in the longer run got hold of the basic principles and were able to use it as a grip to express their ideas. Over time people got used to the line of thinking and were able to put their ideas into the structure. In the end, the structure provided opportunities for the people involved to include their ideas. The central idea in the methodology, 'why should we measure something?' has had great impact.

The overall goal for nature in the Zandmaas/Maasroute protection project was (limited) nature development. This goal was translated into several derived sub-goals. Risks for nature were expected at the interface with other themes. The reduction in river dynamics as a result of the project was considered an important risk to nature.

An overview of the information questions connected to the project goals and derived goals that together form the fundamental objectives, the expected effects, and possible risks for the theme Nature is given in Table 6.1. A cause-effect analysis was included in the information needs. This would enable an improved evaluation of the effects of the Zandmaas/Maasroute monitoring plan on Nature. Socio-economic effects were nevertheless not included. A further elaboration on these information questions and the monitoring aspects is shown in Table 6.2. For each information question in this table it is indicated in what phase of the realization of the Zandmaas/Maasroute protection project monitoring should take place, the method to assess the question, the parameters connected to the question, and the measuring method (the technique or instrument to be used, the (class of) monitoring locations, and the desired resolution (or frequency)) is listed.

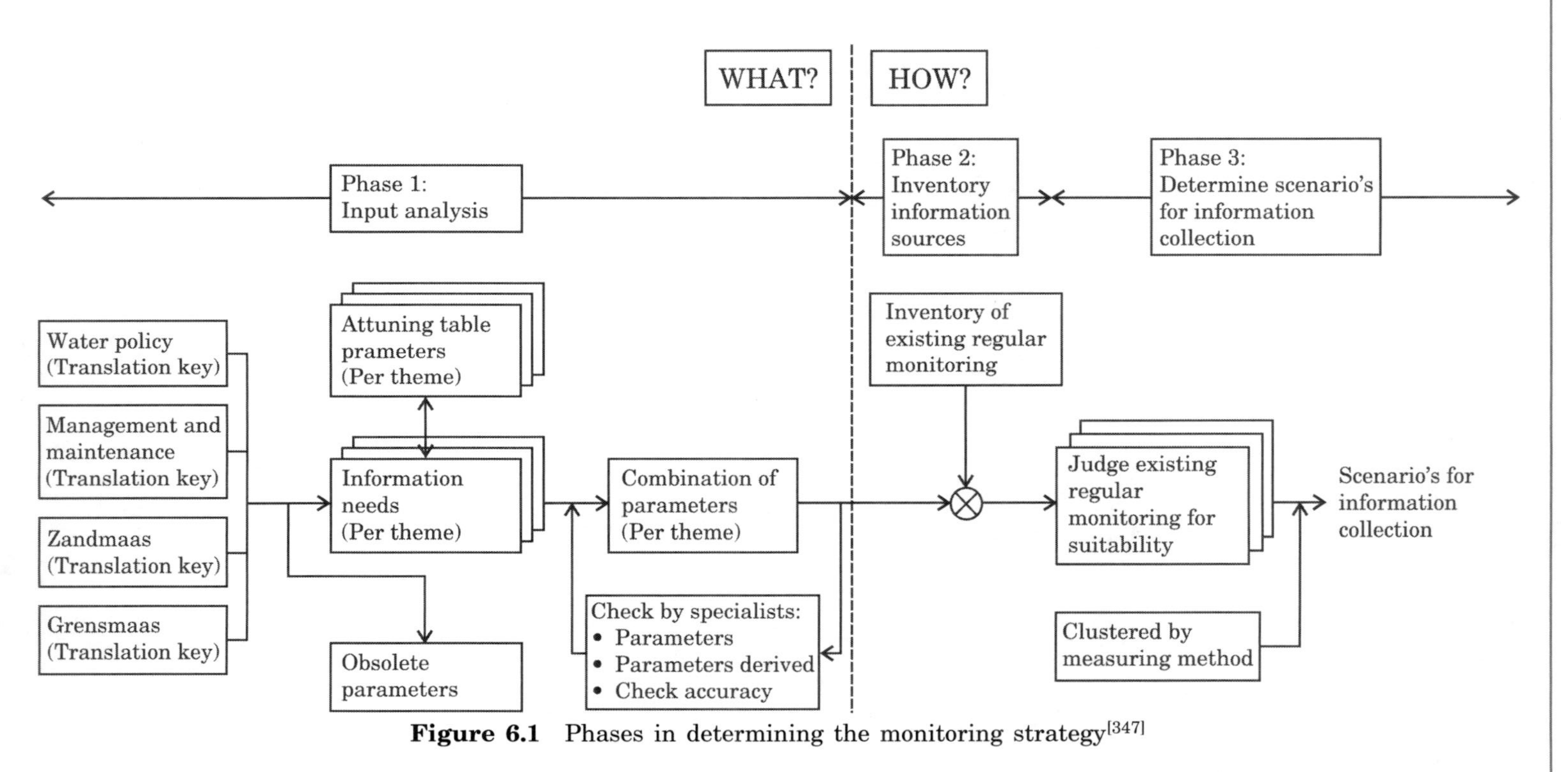

Figure 6.1 Phases in determining the monitoring strategy[347]

Table 6.1 Deduction of information questions from Zandmaas/Maasroute protection project goals as fundamental objectives, effects and risks for the theme Nature (in brackets is the number of the information question as included in Table 6.2)[346]

Project goals	
Realization of limited nature development in conjunction with riverbed widening	What is the surface area and quality of nature developed in conjunction with riverbed widening? (3, 4, 5, 6)
Maintaining existing nature	What existing nature policy categories will be lost or harmed by dredging? (1)
Prevention of (extra) desiccation	Will the desiccation situation change in the vicinity of the river Meuse? (2)
Derived and/or concretized goals:	
Development of river ecotopes	What surface areas of new river ecotopes are developing? (3)
Development of enduring populations of river target species and policy target species	Do possibilities for settlement of viable populations arise based on availability of habitats and spatial coherence of (partial) habitats in the Zandmaas area (6)
Enlargement of surface area and improvement of river ecotopes for river target species	What is the size and quality of habitats for river target species (4, 5)
Prevention of extra eco-toxicological risks for river target species	Do eco-toxicological risks decrease? (7)
Effects	
Destruction of existing nature	What existing nature policy categories will be lost or harmed by dredging? (1)
Change in surface area of river ecotopes	Will 2464 ha of new river ecotopes develop in the predicted ratio? (3)
Habitat availability for river target species	Will the expected habitats for river target species develop in the predicted surface areas? (4)
Habitat quality for river target species	Is the quality of the habitats for river target species in line with the predicted quality? (5)
Network function for river target species	Do possibilities for settlement of viable populations arise based on availability of habitats and spatial coherence of (partial) habitats in the Zandmaas area (6)
Eco-toxicological risks for river target species	Do eco-toxicological risks decrease? (7)
Change in quality of desiccation sensitive nature	Will the desiccation situation change in the vicinity of the river Meuse? (2)
Occurrence of target species policy	Do policy target species occur in the predicted numbers and abundance? (8)
Risks	
Decrease of river dynamics	See habitat availability and habitat quality under effects (4, 5)
Barriers as a result of shipping (Juliana canal)	See network function for river target species under effects (6)

Table 6.2 Overview of information needs in the Zandmaas/Maasroute monitoring project for the theme Nature)[346]

Information question	*Assessment method*	*Parameter*	*Measuring method*		
			Technique/instrument	*Location class*	*Resolution*
1. What existing nature policy categories will be lost or harmed by dredging?	Compare new demarcation of nature areas with existing demarcation	Surface area of specific status and value	Aerial photo analysis	All nature policy categories in the winter bed that will be in contact with dredging areas	Once
2. Will the desiccation situation change in the vicinity of the river Meuse?	Compare present situation with future situation	Phreatophytes Ground water level Ground water quality	Inventory of phreatophytes in permanent test strips Measuring of ground water level Sampling and analysis of ground water	In area sensitive to desiccation outside the winter bed	Phreatophytes once per year in the first 5 years, afterwards once per 3 years Ground water level: see special report
3. Will 2464 ha of new river ecotopes develop in the predicted ratio?	Compare nature, size and spread of ecotopes of predicted to actual situation	Surface area of specific ecotope	Aerial photo analysis	Covering the area	After finalization of the project once per 2 years
4. Will the expected habitats for river target species develop in the predicted surface areas?	Comparison of predicted and measured EOW-index (water system index for ecosystem development) per trajectory	Parameter needed for EOW-index, including: Ecotope distribution Stream velocity Water level	Aerial photo analysis Measuring of water level Measuring of stream velocity	Ecotopes and trajectories	Ecotopes once per 2 years

5. Is the quality of the habitats for river target species in line with the predicted quality?	Comparison of predicted and measured EOW-index per target species	Surface area of specific ecotope Relevance of habitat factors for determining of the habitat quality	Aerial photo analysis Measuring of relevant habitat factors	Covering the area Suitable ecotopes	Ecotopes once per 2 years
6. Are possibilities for settlement of viable populations arising based on availability of habitats and spatial coherence of (partial) habitats in the Zandmaas area?	Analysis of network using LARCH, based on spatial distribution of ecotopes Testing of population size to standards of sustainable populations	Surface area of specific ecotope Target species	Spatial distribution of ecotopes using aerial photo analysis Inventory of species	Covering the area	Once per 2 years
7. Are eco-toxicological risks decreasing for a number of species (groups)?	Comparison of measured concentrations of eco-toxicological substances	PCB-153 Zn Cd Benzo-a-pyrene	Sampling and analyzing parameters in surface water pore water, soil and food of predators	Covering the area	Once per 5 years
8. Are policy target species occurring in the predicted numbers and abundance?	Determine increase and decrease of target species based on present and future situation	Target species	Inventory of target species both in number and spatial distribution		Once per 3 years

In Table 6.2, largely proxies and constructed parameters are used. In the case of desiccation for instance, phreatophytes can be regarded as the natural parameters, as the lack of these plants is the direct problem related to desiccation. Ground water level and quality are proxies that are used to determine if there are other factors involved in desiccation abatement. If ground water level and quality are sufficient, but the vegetation is not restored, other factors should be included in the water policy measures.

The information questions have been recorded for each of the themes that were defined beforehand. The integrated approach requires that the necessary data collection is done in a way that is of use for several assessments and evaluations, across themes. To achieve this goal, in defining the monitoring strategy, links were applied between the measurements per theme and the relevant measurements as recorded in other themes. Efficiency also required that the existing monitoring networks are applied if possible. Three situations were distinguished in this respect:

1. Evaluation can be done using data from existing networks or registers;
2. Evaluation is partly possible using data from existing networks or registers but additional measurements or extending of existing networks is necessary;
3. Evaluation requires an entirely new network.

The results of the prioritization, integration of themes and fitting of existing networks resulted in the overview presented in Table 6.3 for the theme Nature. The table for each measurement identified lists the information question that this measurement should answer the measuring method to be used, the existing monitoring networks, the desire to extend existing monitoring networks, the need to develop new monitoring networks, and other themes that are associated with the measurement and the priority assigned to the measurement.

The various parameters are connected to different information questions, as shown in Table 6.3 and each information question may pose different requirements on the parameter. Table 6.4 lists the requirements for one of the parameters; 'Surface area and location of river ecotopes'. The table clearly shows that mostly the requirements are not too deviant from each other. Still differences in scale, area, and frequency exist and these differences have to be analyzed. Combination of the different requirements in the table will lead to a monitoring network that will provide information for the different information questions. It is clear that in the run of this process, the link with the individual information questions may get lost and thus the link to the original policy objectives. By maintaining a structured approach, the link can be maintained. One way of achieving such a structure is by developing a relational database, which eases maintaining the links, rather than individual tables.

Table 6.3 Monitoring strategy for the integral monitoring plan Zandmaas/Maasroute for the theme Nature[346]. The numbers in the column 'Information question' refer to the number in Table 6.2

Measurements	***Information question***	***Measuring method***	***Existing networks***	***Extend existing networks***	***New networks***	***Associated themes***	***Priority***
Surface area of specific ecotope	1, 3, 5, 6	Aerial photo analysis		Connect to ecotope assessment of the national monitoring network. Extend aerial photos to once per 2 years		Agriculture: 1 Landscape: 3, 4, 5, 8, 9	1
Ground water level	2	Measuring of ground water level	See special report			Landscape: 7 Agriculture: 4	
Phreatophytes in areas sensitive to desiccation	2	Inventory of phreatophytes in permanent test strips		Connect to FLORON, possibly more emphasis on phreatophytes Eco-hydrological atlas once per 3 years			1
Stream velocity	4	Ott mills, ADM (acoustic discharge meter)			Network stream velocity	Safety: 1, 2, 3, 4, 9, 10, 11, 12 Shipping: 9	2

Table 6.3 (*Contd...*)

Table 6.3 (*Contd...*) Monitoring strategy for the integral monitoring plan Zandmaas/Maasroute for the theme Nature[346]. The numbers in the column 'Information question' refer to the number in Table 6.2

Measurements	*Information question*	*Measuring method*	*Existing networks*	*Extend existing networks*	*New networks*	*Associated themes*	*Priority*
Water level	4	Automatic level detection		Extend water level measurement to one location for each trajectory		Safety: 1, 2, 3, 4, 9, 10, 11, 12 Shipping: 1, 3, 7 Landscape: 4 Agriculture: 4 Living, working and infrastructure: 3	2
Parameter needed for EOW-index	4	?	?	?	?	?	2
Relevant habitat factors for the determination of the habitat quality	5	Measuring of relevant habitat factors	?	?	?	?	2
Target species	6, 8	Inventory of species (number and spatial distribution)	Inventory of species through PGO's (private data-managing organizations)				1

PCB-153 Zn Cd Benzo-a-pyrene	7	Sampling and analyzing parameters in surface water pore water, soil and food of predators	Extend present chemical network with extra locations	Network for parameters in pore water, soil and food of predators	Water quality and sludge: 2, 3	2

Table 6.4 Overview of frequencies and locations for the parameter 'Surface area and location of river ecotopes'[341]

Accuracy	***Method/ technique wanted***	***Operational characteristics wanted***			***Method/ technique chosen***	***Operational characteristics chosen***					
		Phase	***Frequency***	***Location***		***Frequency chosen***	***Frequency existing***	***Locations chosen***	***Total locations***	***No. new***	***No. existing***
Scale 1:10.000 $m^2 \pm 10\%$	Inventory with use of aerial photo mapping.	Permanent	Once per 4 years	Covering the area and surroundings	Aerial photo mapping (true color)	Once per 4 years	Once per 8 years	Covering the river Meuse area	224	0	224
Scale 1:5.000 $m^2 \pm 10\%$	Additional field exploration	Permanent	Once per 4 years	Covering the area and surroundings	Field inventory	Once per 4 years	None	Covering the river Meuse area	224	224	0
Scale 1:10.000 $m^2 \pm 10\%$	Inventory with use of aerial photo mapping	Permanent	Once per 8 years	Covering the area and surroundings	Aerial photo mapping (true color)	Once per 8 years	None	Covering the area of the Meuse canals and Central Limburg canals	260	260	0

Scale 1:5.000 $m^2 \pm 10\%$	Aerial photo mapping	Realization, Management	Zero situation, once after 1, 4, 8 years after realization subsequently once in 8 years	Covering the area	Aerial photo mapping	Once after 1, 4, 8 years after realization subsequently once per 8 years	Once per 8 years	Covering the area	41	0	41
Scale 1:5.000 $m^2 \pm 10\%$	Additional field exploration	Permanent	Zero situation, once after 1, 4, 8 years after realization subsequently once in 8 years	Covering the area	Additional field exploration	Once after 1, 4, 8 years after realization subsequently once per 8 years	Once per 8 years	Covering the area	41	0	41
Scale 1:5.000 $m^2 \pm 10\%$	Aerial photo interpretation Additional field exploration	Present, management	Once per 2 years	Covering the Zandmaas area—winter bed	Aerial photo interpretation Additional field exploration	Once after 1, 4, 8 years after realization subsequently once per 8 years	Once per 8 years	Covering the Zandmaas area	159	0	159

Note that the information needs were focused on evaluation of the program of measures, rather than on evaluation of the project as a whole. Consequently, in this structured breakdown of the problem, the socio-economic aspects were largely neglected. Evaluation of that part of the project after finalization was as a result unsatisfactory. Also here, the purpose of evaluation of the project leads to emphasis on information on the effectiveness of measures rather than on evaluation of the overall project goals. Through the approach of looking at goals, problems (or risks) and measures (effects), this bias seems to be encountered to some extent. It is nevertheless clear that the selection of objectives has a large influence on the outcomes of the process.

The overall project consisted of four major projects of which the Zandmaas/Maasroute was one. Each project independently specified information needs and developed a monitoring plan. After finalization of the monitoring plans, the first step was to integrate the four plans. To be able to do this, sometimes a reconsideration of the information needs was needed, because it was not always clear enough if parameters from different plans could be considered as equal to each other. Also, during the phases of integration and development of data collection scenario's it appeared that especially the required accuracy of parameters was not specified well enough to determine if the frequencies and locations would cover for this. From these examples it becomes clear that the information cycle, although it is portrayed as a linear process, cannot be passed through in one direction but needs regular reconsideration of previous steps. As such it can never be a linear process.

Even after merging of the four monitoring plans into a monitoring strategy, they remained clearly distinguishable, uninfluenced by the others. Through this broad approach, a comprehensive overview of information needs from the four different viewpoints was achieved, enabling attuning of the measuring needs and setting of priorities between the different aspects. Further advantages of the structured approach through the rugby ball methodology were[340]:

- The need for monitoring could be indicated with a high level of detail. This prevents discourse over the importance of different components and allows a discussion concerning the content.
- The four monitoring plans have been elaborated in a uniform way on the basis of the method. This has simplified harmonization of the four plans.
- The step-by-step specification of information needs facilitates a complete end product. If the process goes too fast, one runs the risk that only few participants can provide input and that the end product will be incomplete.

As the amount of information collected in this study was huge, storing and presentation of the information must be well thought over in advance. In the study, the information was reported through extensive tables. Although these tables were well developed, the volume of information made it difficult to access. Use of a database from the beginning onwards can avoid such problems. From the evaluation of the overall project it was concluded that the strongest improvements of the study as related to the usual approach of monitoring network design is formed by[340]:

- The broad approach through which all themes have been incorporated from the various viewpoints of the information users;
- The structured approach and reporting through which the monitoring process is made transparent.

6.6 EXERCISES

- Use the information collected in the previous chapters to write the information needs report. To this end, use the template for the report as described in Box 6.1. Also identify if any information is missing to complete the template. If this is the case, turn back to the relevant chapter and identify which omissions are made while performing the exercises in that chapter that have led to information missing in the end of the process.
- Based on the assessment of information needs as performed in the previous chapter, develop an overview of the measurable attributes and the related requirements for each attribute. Identify the various attributes as specified and make a listing with reference to the respective policy objectives.
- Identify similar attributes from different policy objectives and compare the various requirements. If the requirements are missing or ambiguous, go back to the exercises in Chapter 5 and complete the missing information.
- Analyze the existing information networks and determine what information is collected that can provide for the information needs. Compare the requirements from the inventory to the specifications of the information networks and determine of these specifications meet the requirements.
- From the analysis, develop a monitoring strategy including the attributes on which data is to be collected, the sources of the data, method of data collection to be used, measuring locations, and frequencies.

References

[1] Cosgrove, W.J., Rijsberman, F.R. (2000) *World water vision; making water everybody's business*. London, UK: Earthscan Publications Ltd.

[2] Hinrichsen, D. (2003) *A human thirst*. World Watch Magazine. Washington DC, USA: Worldwatch Institute. p. 12-8.

[3] Saeijs, H.L.F., van Berkel, M.J. (1995) Global water crisis: the major issue of the 21st century, a growing and explosive problem. *European Water Pollution Control* 5(4):26-40.

[4] Serageldin, I., Steer, A. (1994) Epilogue: expanding the capital stock. *Making development sustainable: from concepts to actions*. Washington D.C., USA: The World Bank. p. 30-2.

[5] GWP-TAC (2000) *Integrated Water Resources Management*. Stockholm, Sweden: Global Water Partnership - Technical Advisory Committee.

[6] UN-WWAP (2006) *Water, a shared responsibility. The United Nations World Water Development report 2*: UNESCO Publishing.

[7] UN-WWAP (2003) *Water for people, Water for life. The United Nations World Water Development report*. Barcelona, Spain: UNESCO Publishing.

[8] Rauch, W. (1998) Problems of decision making for a sustainable development. *Water Science and Technology* 38(11):31-9.

[9] Allan, T. (2003) *IWRM/IWRAM: a new sanctioned discourse?* London, UK: University of London.

[10] Biswas, A.K. (2004) Integrated water resources management: a reassessment. A water forum contribution. *Water International* 29 (2):248-56

[11] Kabat, P., van Schaik, H. (2003) *Climate changes the water rules: How water managers can cope with today's climate variability and tomorrow's climate change.*

[12] Falkenmark, M. (2000) *No freshwater security without major shift in thinking. Ten-year message from the Stockholm Water Symposia.* Stockholm, Sweden: Stockholm International Water Institute, SIWI.

[13] Boyle, M., Kay, J., Pond, B., Munn, T. (2001) Monitoring in support of policy: an adaptive ecosystem approach. *Encyclopedia of global environmental change, Volume 4*: John Wiley and Son. p. 116-37.

[14] Dinar, A. (1998) Water policy reforms: information needs and implementation obstacles. *Water Policy* 1:367-82.

[15] Gouveia, C., Fonseca, A., Camara, A., Ferreira, F. (2004) Promoting the use of environmental data collected by concerned citizens through information and communication technologies. *Journal of Environmental Management* 71:135.

[16] Healy, R.G., Asher, W. (1995) Knowledge in the policy process: Incorporating new environmental information in natural resources policy making. *Policy Sciences* 28:1-19.

[17] Allen, W., Bosch, O., Kilvington, M., Brown, I., Harley, D. (2001) Monitoring and adaptive management: resolving social and organizational issues to improve information sharing in natural resource management. *Natural Resources Forum* 25(3):225-33.

[18] Haklay, M.E. (2003) Public access to environmental information: past, present and future. *Computers, Environment and Urban Systems* 27:163-80.

[19] Lovett, G.M., Burns, D.A., Driscoll, C.T., Jenkins, J.C., Mitchell, M.J., Rustad, L., et al. (2007) Who needs environmental monitoring? *Frontiers in Ecology and the Environment* 5:253-60.

[20] Perry, J.A., Vanderklein, E. (1996) *Water Quality: Management of a natural resource.* Cambridge, MA, USA: Blackwell Science.

[21] Meybeck, M. Surface water quality: global assessment and perspectives In: Zebidi, H. editor. *Conference Surface water quality: global assessment and perspectives.* p. 173-85.

[22] Meybeck, M. (2004) The global change of continental aquatic systems: dominant impacts of human activities. *Water Science & Technology* 49 (7):73-83.

[23] Meybeck, M., Helmer, R. (1989) The quality of rivers: from pristine state to global pollution. 75:283-309.

[24] Groshart, C.P., Wassenberg, W.B.A., Laane, R.W.P.M. (2000) *Chemical study on brominated flame-retardants.* The Hague, Netherlands: RIKZ.

[25] Jongbloed, R.H., Visschedijk, A.J.H., van Dokkum, H.P., Laane, R.W.P.M. (2000) *Toxaphene. An analysis of possible problems in the aquatic environment*. The Hague, Netherlands: RIKZ.

[26] Vethaak, A.D., Rijs, G.B.J., Schrap, S.M., Ruiter, H., Gerritsen, A., Lahr, J. (2002) *Estrogens and xeno-estrogens in the aquatic environment of the Netherlands. Occurrence, potency and biological effects*. The Hague/Lelystad, Netherlands: RIZA/RIKZ.

[27] Lorenz, C.M., Gilbert, A.J., Cofino, W.P. (2001) Environmental auditing: Indicators for transboundary river management. *Environmental Management* 28(1):115-29.

[28] van Kerkhoff, L. (2005) Integrated research: concepts of connection in environmental science and policy. *Environmental Science and Policy* 8:452-63.

[29] Bruch, C.E. (2000) *Comparative policy and practice of access to environmental information*. Dublin, Ireland.

[30] Commission, E. (2003) Directive 2003/4/EC of the European Parliament and of the Council of 28 January 2003 on public access to environmental information and repealing Directive 90/313/EEC of the council. *Official Journal of the European Communities* 14.2.2003:L 41/26-L 41/32.

[31] UNCED (1992) *Agenda 21, the Rio Declaration on Environment and Development*. Rio de Janerio, Brazil: United Nations Conference on Environment and Development, 3 to 14 June 1992.

[32] UNECE (1998) *Convention on access to information, public participation in decision-making and access to justice in environmental matters*. Aarhus, Denmark: UN Economic Commission for Europe.

[33] Anonymous (1969) *National Environmental Policy Act of 1969*

[34] de Villeneuve, C.H.V., Timmerman, J.G., Langaas, S. (2004) Legal aspects of information in transboundary river basin management. *Environmental information in European transboundary water management*. London, UK: IWA Publishing. p. 60-77.

[35] WCMC (1998) *Volume 1: Information and policy*. London, UK: World Conservation Monitoring Centre. Commonwealth secretariat.

[36] Harremoës, P., Gee, D., MacGarvin, M., Stirling, A., Keys, J., Wynne, B., et al. (2001) *Late lessons from early warnings: the precautionary principle 1896-2000*. Copenhagen, Denmark.

[37] Denisov, N., Christoffersen, L. (2001) *Impact of environmental information on decision making processes and the environment*. Arendal, Norway.

[38] van Bracht, M.J. (2001) *Made to measure. Information requirements and groundwater level monitoring networks* [Doctoral thesis]: Vrije Unversiteit Amsterdam.

[39] van Bracht, M.J. (1997) Expected damage. A surrogate measure for quantifying the advantages of groundwater monitoring networks for water management In: Ottens, J.J., Claessen, F.A.M., Stoks, P.G., Timmerman, J.G., Ward, R.C. editors. *Conference Expected damage. A surrogate measure for quantifying the advantages of groundwater monitoring networks for water management*, Lelystad, The Netherlands. RIZA, p. 129-34.

[40] Markowitz, A. Communicating monitoring results that people can understand. *Conference Communicating monitoring results that people can understand*, Austin, Texas, USA. p. 50-4.

[41] Timmerman, J.G., Cofino, W.P. (2001) Main findings of the international workshop Monitoring Tailor-Made III - Information for sustainable water management In: Timmerman, J.G., Cofino, W.P., Enderlein, R.E., J lich, W., Literathy, P.L., Martin, J.M., et al. editors. *Conference Main findings of the international workshop Monitoring Tailor-Made III - Information for sustainable water management*, Lelystad, The Netherlands. RIZA / IWAC, p. 395-9.

[42] Davenport, T.H. (1997) *Information ecology; Mastering the information and knowledge environment*. New York: Oxford University Press.

[43] Russell, C.S. (2001) Monitoring, enforcement, and the choice of environmental policy instruments. *Regional Environmental Change* 2(2):73-6.

[44] Breukel, R.M.A. (2000) *Nation-wide water monitoring: the experience of the Netherlands. Presentation at the OECD/SEPA seminar on environmental monitoring Beijing, China. April 12-14, 1999*. Lelystad, The Netherlands: RIZA.

[45] Warmer, H., van Dokkum, R. (2002) *Water pollution control in the Netherlands. Policy and practice 2001*. Lelystad, The Netherlands.

[46] Bosch, A., van der Ham, W. (1998) *Twee eeuwen Rijkwaterstaat (Two ages of Rijkswaterstaat) 1798-1998*. Europese bibliotheek, Zaltbommel: Rijkswaterstaat (Den Haag) / Stichting Historie der Techniek (Eindhoven).

[47] RIZA (1965) *Jaarverslag 1964 (Yearreport 1964)*. The Hague, The Netherlands: Staatsuitgeverij.

[48] Timmerman, J.G., Buiteveld, H., Lamers, M., Möllenkamp, S., Isendahl, N., Ottow, B., et al. (2010) Case Study: Rhine. In: Mysiak, J., Henriksen, H.J., Sullivan, C., Bromley, J., Pahl-Wostl, C., editors. *The adaptive Water resource Management Handbook*. London, UK: Earthscan. p. 117–28

[49] Huisman, P. (1996) Water management in the Rhine delta. *Conference Water management in the Rhine delta*, Koblenz, germany. Deutsches IHP/OHP-Nationalkomitee, p. 79-96.

[50] Anonymous (1968) *De waterhuishouding van Nederland (the water management of The Netherlands)*. 's-Gravenhage, The Netherlands: Staatsuitgeverij.

[51] Anonymous (1970) Besluit van 5 november 1970, ter uitvoering van de Wet verontreiniging oppervlaktewateren met betrekking tot oppervlaktewateren onder beheer van het Rijk en de volle zee (Decision of 5 November 1970 for the implementation of the Act on pollution of surface waters regarding surface water under the National responsibility and the wider sea). *Staatsblad* 377.

[52] Anonymous (1983) Besluit van 3 november 1983 houdende regelen inzake kwaliteitsdoelstellingen en metingen oppervlaktewateren (Decision of 3 November 1983 on regulation concerning quality standards and measurements of surface waters). *Staatsblad* 606.

[53] Anonymous (1984) *De waterhuishouding van Nederland : 1984 (tweede nota waterhuishouding)*. Ministerie van Verkeer en Waterstaat, 's- Gravenhage.

[54] European_Commission (1975) Council Directive of 16 June 1975 concerning the quality required of surface water intended for the abstraction of drinking water in the Member States (75/440/EEC). *Official Journal of the European Communities* 25.7.75:L 194/26-L /31.

[55] European_Commission (1976) Council Directive of 8 December 1975 concerning the quality of bathing water (76/160/EEC). *Official Journal of the European Communities* 5.2.76:L 31/1-L /7.

[56] European_Commission (1978) Council Directive of 18 July 1978 on the quality of fresh waters needing protection or improvement in order to support fish life (78/659/EEC). *Official Journal of the European Communities* 14.8.78:L 222/1-L /10.

[57] European_Commission (1979) Council Directive of 30 October 1979 on the quality required of shellfish waters (79/923/EEC). *Official Journal of the European Communities* 10.11.79:L 281/47-L /52.

[58] Timmerman, J.G., Houben, A., van Steenwijk, J., van der Weijden, M. (2004) *Legal obligations for Dutch national water quality monitoring*. Lelystad, Netherlands: RIZA.

[59] Anonymous (1999) *New deal with an old enemy. Water management in the Netherlands: Past, present and future*. Lelystad, The Netherlands: Ministry of Transport, Public Works and Water Management.

[60] Anonymous (1998) *Fourth national policy document on water management; Government decision (abridged version)* The Hague, The Netherlands: Ministry of Transport, Public Works and Water Management.

[61] Anonymous (2009) *2009-2015 National Water Plan—A Summary*. The Hague, The Netherlands: State Secretary for Transport,

Public Works and Water Management and the Ministers for Housing, Regional Development and the Environment and for Agriculture, Nature and Food Quality.

[62] van Ast, J.A. (2003) Towards interactive water management in international river basins In: Lombardo, C., Coenen, M., Sacile, R., Meire, P. editors. *Conference Towards interactive water management in international river basins*, Antwerp, Belgium. NATO Committee on the Challenges of Modern Society, p. 77-80.

[63] Allan, C., Xia, J., Pahl-Wostl, C. (2013) Climate change and water security: challenges for adaptive water management. *Current Opinion in Environmental Sustainability* 5(6):625-32.

[64] van der Brugge, R., Rotmans, J., Loorbach, D. (2004) The transition in Dutch water management. *Regional Environmental Change* 5(4):164-76.

[65] Anonymous (1990) *Water voor nu en later; Derde nota waterhuishouding. Regeringsbeslissing. (Water for now and the future; Third national policy document on water management. Government decision)*. The Hague, The Netherlands.

[66] Anonymous (1991) *Water in the Netherlands: a time for action*. The Hague, The Netherlands: Ministry of Transport, Public Works and Water Management.

[67] Blom, G. Dutch water management in practice, the REGIWA-projects: experiences, results, future developments In: Romijn, E., Sch tte de Roon, J.C. editors. *Conference Dutch water management in practice, the REGIWA-projects: experiences, results, future developments*, Lelystad, The Netherlands. RIZA, p. 3-12.

[68] Wisserhof, J. Enhancing research utilization for integrated water management. *Conference Enhancing research utilization for integrated water management*, Amsterdam, The Netherlands. IAWQ / EWPCA / NVA, p. 431-40.

[69] de Jong, J., van Rooy, P.T.J.C., Hosper, S.H. (1994) Living with water: at the cross-roads of change. *Conference Living with water: at the cross-roads of change*, Amsterdam, The Netherlands. IAWQ/EWPCA/NVA, p. 477-93.

[70] Engelen, G.B., Kloosterman, F.H. (1996) *Hydrological systems analysis: methods and application*. Dordrecht, The Netherlands: Kluwer Academic Pub.

[71] Raskin, P., Gleick, P., Kirshen, P., Pontius, G., Strzepek, K. (1997) *Water future: Assessment of long-range patterns and problems*. Stockholm, Sweden: Stockholm Environment Institute.

[72] Somlyódy, L. (1994) Water quality management: can we improve integration to face future problems? *Conference Water quality management: can we improve integration to face future problems?* 1994, p. 343-59.

[73] van Rooy, P.T.J.C., de Jong, J. (1995) Towards comprehensive water management in the Netherlands: (1) developments. *European Water Pollution Control* 5(4):59-66.

[74] Anonymous (1998) *Waterkader; Vierde nota waterhuishouding. Regeringsbeslissing. (Water frame; Fourth national policy document on water management. Government decision).* The Hague, The Netherlands: Ando bv.

[75] Anonymous (2000) *Ministerial declaration of The Hague on water security in the 21st century.* The Hague, The Netherlands: Second World Water Forum.

[76] Teunissen, K. (2004) *Initiatieven bewoners Overdiepse polder beloond (Initiatives of inhabitants Overdiepse polder rewarded).* de Waterp. 7-8.

[77] EC (2000) Directive 2000/60/EC of the European Parliament and of the Council of 23 October 2000 establishing a framework for Community action in the field of water policy. *Official Journal of the European Communities* 22.12.2000:L 327/1-L /72.

[78] Timmerman, J.G. (2011) *Bridging the water information gap : structuring the process of specification of information needs in water management* [Proefschrift Wageningen Met lit. opg.—Met samenvatting in het Engels en Nederlands Auteursvermelding op omslag: Jos Timmerman]. Wageningen, The Netherlands: Wageningen University.

[79] Groot, S. (1981) *Het optimaliseren en struktureren van het meetnet van de kwaliteit der Rijkswateren (OSTWAT): beschouwing en toepassing van de methode Lettenmaier: verslag onderzoek. (Optimising and structuring of the monitoring network for the quality of the national waters (OSTWAT): discussion and application of the method Lettenmaier: study report).* Delft, The Netherlands: Delft Hydraulics.

[80] RIZA (1972) *Jaarboek der waterkwaliteit van de Rijkswateren 1965; Rijntakken en Maas. (Yearbook of the water quality of the national waters 1965; Rhine and its branches and Meuse).* The Hague, The Netherlands: Staatsuitgeverij.

[81] RIZA (1972) *Jaarboek der waterkwaliteit van de Rijkswateren 1966; Rijntakken en Maas. (Yearbook of the water quality of the national waters 1966; Rhine and its branches and Meuse).* The Hague, The Netherlands: Staatsuitgeverij.

[82] RIZA (1967) *Jaarverslag 1966 (Yearreport 1966).* The Hague, The Netherlands: Staatsuitgeverij.

[83] Timmerman, J.G., Beinat, E., Termeer, C.J.A.M., Cofino, W.P. (2010) Analysing the data-rich-but-information-poor syndrome in Dutch water management in historical perspective. *Environmental Management* 45(5):1231-42.

[84] Chave, P.A. (2001) *The EU Water Framework Directive. An introduction*. London, UK: IWA Publishing.

[85] de Jong, J., van Buuren, J.T., Luiten, J.P.A. (1996) Systematic approaches in water management: aquatic outlook and decision support systems combining monitoring, research, policy analysis and information technology. *Water Science and Technology* 34(12):9-16.

[86] Lindenmayer, D.B., Likens, G.E. (2009) Adaptive monitoring: a new paradigm for long-term research and monitoring. *Trends in Ecology and Evolution* 24(9):482-6.

[87] Vaes, G., Willems, P., Swartenbroekx, P., Kramer, K., de Lange, W., Kober, K. (2009) Science-policy interfacing in support of the Water Framework Directive implementation. *Water Science and Technology* 60(1):47-54.

[88] Wesselink, A., de Vriend, H., Barneveld, H., Krol, M., Bijker, W. (2009) Hydrology and hydraulics expertise in participatory processes for climate change adaptation in the Dutch Meuse. *Water Science and Technology* 60(3):583-95.

[89] Strobl, R.O., Robillard, P.D. (2008) Network design for water quality monitoring of surface freshwaters: A review. *Journal of Environmental Management* 87:639-48.

[90] Mancy, K.H., Allen, H.E., Suess, M.J. (1982) Design of measurement systems. Oxford, UK: WHO/Pergamon press.

[91] Steele, T.D. (1987) Water quality monitoring strategies. *Hydrological Sciences Journal* 32(2):207-13.

[92] Ward, R.C., Loftis, J.C., McBride, G.B. (1986) The "Data-rich but Information-poor" syndrome in water quality monitoring. *Environmental Management* 10(3):291-7.

[93] Ward, R.C. (1996) Water quality monitoring: where 's the beef? *Water Resources Bulletin* 32(4):673-80.

[94] Gooch, G.D., Stalnacke, P. (2006) *Integrated transboundary water management in theory and practice: Experiences from the new EU Eastern borders*. London, UK: IWA Publishing.

[95] Greeuw, S.C.H., van Asselt, M.B.A., Grosskurth, J., Storms, C.A.M.H., Rijkens-Klomp, N., Rothman, D.S., et al. (2000) *Cloudy crystal balls. An assessment of recent European and global scenario studies and models*. Copenhagen, Denmark: European Environment Agency.

[96] Hisschemöller, M., Tol, R.S.J., Vellinga, P. (2001) The relevance of participatory approaches in integrated environmental assessment. *Integrated Assessment* 2:57-72.

[97] Timmerman, J.G., Cofino, W.P. (2003) Main findings In: Timmerman, J.G., Behrens, H.W.A., Bernardini, F., Daler, D., Ross,

P., Ruiten, C.J.M., et al. editors. *Conference Main findings*, Lelystad, The Netherlands. RIZA / IWAC / RIKZ, p. 333-8.

[98] Scheuer, S. (2005) *EU Environmental Policy Handbook. A critical analysis of EU environmental legislation. Making it accessible to environmentalists and decision makers.* Brussels | Belgium European Environmental Bureau (EEB)

[99] Bauer, M. (2007) Get valuable information from the data graveyard. *South African Journal of Industrial Engineering* 18(1):145-56.

[100] Blake, J. (1999) Overcoming the 'Value-Action Gap' in environmental policy: tensions between national policy and local experience. *Local Environment* 4(3):257-78.

[101] McNie, E.C. (2007) Reconciling the supply of scientific information with user demands: an analysis of the problem and review of the literature. *Environmental Science and Policy* 10:17-38.

[102] Cash, D.W., Clark, W.C., Alcock, F., Dickson, N.M., Eckley, N., Guston, D.H., et al. (2003) Knowledge systems for sustainable development. *Proceedings of the National Academy of Science* 100(14):8086-91.

[103] Hollick, M. (1981) The role of quantitative decision-making methods in environmental impact assessment. *Journal of Environmental Management* 12:65-78.

[104] Bemelmans, T.M.A. (1989) *Bestuurlijke informatiesystemen en automatisering (Managerial information systems and computerisation)* Leiden, The Netherlands: Stenfert Kroese.

[105] de Leeuw, A.C.J. (1988) *Organisaties: management, analyse, ontwerp en verandering: een systeemvisie. (Organisations: management, analysis, design and change: a systems vision)* Assen, The Netherlands: Van Gorkum.

[106] Ehin, P. (2003) *Theoretical approaches to public participation.* Tartu, Estonia.

[107] Boogerd, A., Groenewegen, P., Hisschem"ller, M. (1997) Knowledge utilization in water management in the Netherlands related to desiccation. *Journal of the American Water Resources Association* 33(4):731-40.

[108] Burgess, J., Harrison, C.M., Filius, P. (1998) Environmental communication and the cultural politics of environmental citizenship. *Environment and Planning* 30:1445-60.

[109] Hofstra, M.A. (1995) Information is vital for the national decision maker In: Adriaanse, M., van de Kraats, J., Stoks, P.G., Ward, R.C. editors. *Conference Information is vital for the national decision maker*, Lelystad, The Netherlands. RIZA, p. 43-54.

[110] Timmerman, J.G., Ottens, J.J. (1997) The information cycle - a framework for the management of our water resources In: Siwi

editor. *Conference The information cycle - a framework for the management of our water resources*, Stockholm, Sweden. p. 407-15.

[111] Bradshaw, G.A., Borchers, J.F. (2000) Uncertainty as information: narrowing the science-policy gap. *Conservation Ecology* 4(1):7.

[112 Milich, L., Varady, R.G. (1999) Openness, sustainability, and public participation: new designs for transboundary river basin institutions. *Journal of Environment & Development* 8(3):258-306.

[113] Adriaanse, M., van de Kraats, J., Stoks, P.G., Ward, R.C. (1995) Conclusions Monitoring Tailor-Made In: Adriaanse, M., van de Kraats, J., Stoks, P.G., Ward, R.C. editors. *Conference Conclusions Monitoring Tailor-Made*, Beekbergen, The Netherlands. RIZA, p. 345-7.

[114] Ongley, E.D. (1998) Modernisation of water quality programmes in developing countries: Issues of relevancy and cost efficiency. *Water Quality International* September/October:37-42.

[115] Ten Brink, B.J.E., Woudstra, J.H. (1991) Towards an effective and rational water management: The Aquatic Outlook Project - integrating water management, monitoring and research. *European Water Pollution Control* 1(5):20-7.

[116] van Dooren, W. (2004) Supply and demand of policy indicators. A cross-sectoral comparison. *Public Management Review* 6(4):511-30.

[117] Ward, R.C. (1995) Monitoring Tailor-Made: What do you want to know? In: Adriaanse, M., van de Kraats, J., Stoks, P.G., Ward, R.C. editors. *Conference Monitoring Tailor-Made: What do you want to know?*, Lelystad, The Netherlands. RIZA, p. 16-24.

[118] Adriaanse, M. Tailor-made guidelines: a contradiction in terms? In: Ottens, J.J., Claessen, F.A.M., Stoks, P.G., Timmerman, J.G., Ward, R.C. editors. *Conference Tailor-made guidelines: a contradiction in terms?*, Nunspeet, The Netherlands. RIZA, p. 391-9.

[119] Bartram, J., Helmer, R., Ballance, R. (1996) Introduction. *Water quality monitoring - A practical guide to the design and implementation of freshwater quality studies and monitoring programmes*. London, UK: Chapman & Hall. p. 1-8.

[120] Meybeck, M., Kimstach, V., Helmer, R., Chapman, D. (1996) Strategies for water quality assessment. *Water quality assessment - A guide to the use of biota, sediments and water in environmental monitoring*. London, UK: Chapman & Hall. p. 19-50.

[121] Adriaanse, M., Lindgaard-J›rgensen, P., Helmer, R., Hespanol, I. (1997) Information systems. *Water Pollution Control A guide to the use of water quality management principles*. London, UK: UNEP/E&FN Spon. p. 245-74.

[122] Bernstein, B.B., Hoenicke, R., Tyrell, C. (1997) Planning tools for developing comprehensive regional monitoring programs. *Environmental Monitoring and Assessment* 48:297-306.

[123] Brett, M. (2000) *Environmental information: knowledge management of supply and demand. Part 1: Estimating the magnitude for the demand of environmental information*. Dublin, Ireland.

[124] MacDonald, L.H. (1994) Developing a monitoring project. *Journal of Soil and Water Conservation* May-June 1994:221-7.

[125] Ouwersloot, J. (1994) *Information and communication from an economic perspective. Conceptual models and empirical analysis* [Ph.D.]: Free University, Amsterdam, The Netherlands.

[126] USGS (1994) *Water quality monitoring in the United States - Technical appendixes. 1993 report of the intergovernmental task force on monitoring water quality*. Washington D.C., USA: U.S. geological Survey.

[127] van Luin, A.B., Ottens, J.J. (1997) Conclusions and recommendations In: Ottens, J.J., Claessen, F.A.M., Stoks, P.G., Timmerman, J.G., Ward, R.C. editors. *Conference Conclusions and recommendations*, Lelystad, The Netherlands. RIZA, p. 401-3.

[128] Ward, R.C., Loftis, J.C., McBride, G.B. (1990) *Design of water quality monitoring systems*. New York, USA: Van Nostrand Reinhold.

[129] Beinat, E. (1995) *Multiattribute value functions for environmental management* [Ph.D.]: Free University, Amsterdam, The Netherlands.

[130] Dunn, W.N. (1994) *Public policy analysis: an introduction*. New Jersey, USA: Prentice-Hall.

[131] WWAP (2012) *The United Nations World Water Development Report 4: Managing Water under Uncertainty and Risk*. World Water Assessment Programme France: UNESCO-WWAP.

[132] Lahdelma, R., Salminen, P., Hokkanen, J. (2000) Using multi-criteria methods in environmental planning and management. *Environmental Management* 26(6):595-605.

[133] Cash, D.W., Moser, S.C. (1998) Information and decision making systems for the effective management of cross-scale environmental problems: A theoretical concept paper. *Conference Information and decision making systems for the effective management of cross-scale environmental problems: A theoretical concept paper*, Cambridge, MA. Harvard University.

[134] Guedes Vaz, S., Martin, J., Wilkinson, D., Newcombe, J. (2001) *Reporting on environmental measures: are we being effective?* Copenhagen, Denmark: European Environment Agency.

[135] GWP-TEC (2004) *Catalyzing Change: A handbook for developing integrated water resources management (IWRM) and water efficiency strategies*. Stockholm, Sweden Global Water Partnership - Technical Committee.

[136] Holsapple, C.W., Moskowitz, H. (1980) A conceptual framework for studying complex decision processes. *Policy Sciences* 12:83-104.

[137] Kliot, N., Shmueli, D. (2001) Development of institutional frameworks for the management of transboundary water resources. *International Journal of Global Environmental Issues* 1(3/4):306-28.

[138] Musters, C.J.M., de Graaf, H.J., Ter Keurs, W.J. (1998) Defining socio-environmental systems for sustainable development. *Ecological Economics* 26:243-58.

[139] van Roost, M., Ruijgh-Van der Ploeg, M. Indicators for water polict evaluation from a network management perspective In: Timmerman, J.G., Cofino, W.P., Enderlein, R.E., J lich, W., Literathy, P.L., Martin, J.M., et al. editors. *Conference Indicators for water polict evaluation from a network management perspective*, Lelystad, The Netherlands. RIZA / IWAC, p. 177-84.

[140] Winsemius, P. (1986) *Gast in eigen huis, beschouwingen over milieumanagement (Guest in own home, contemplating on environmental management)*. Alphen a/d Rijn, The Netherlands: Samson H.D. Tjeenk Willink.

[141] Ingram, H. Science and environmental policy. *Conference Science and environmental policy*, vol. Annual Meeting of the Pacific Division of the Association for the Advancement of Science June 19, 2001.

[142] Rittel, H.W.J., Webber, M.M. (1973) Dilemmas in a general theory of planning. *Policy Sciences* 4:155-69.

[143] WRR (2003) *Naar nieuwe wegen in het milieubeleid (Towards new roads in environmental policy)* Den Haag: SDU Uitgevers.

[144] Buckingham Shum, S. Representing hard-to-formalise, contextualised, multidisciplinary, organisational knowledge. *Conference Representing hard-to-formalise, contextualised, multidisciplinary, organisational knowledge*, Palo Alto, CA, USA. Stanford University, AAAI Press.

[145] Klijn, E.H., van Bueren, E., Koppejan, J.M.F. (2000) *Spelen met onzekerheid: over diffuse besluitvorming in beleidsnetwerken en mogelijkheden voor management*. Delft, The Netherlands: Eburon.

[146] Funtowicz, S.O., Martinez-Alier, J., Munda, G., Ravetz, J.R. (1999) *Information tools for environmental policy under conditions of complexity*. Copenhagen, Denmark: European Environment Agency.

[147] Imo (2002) *Anti-fouling systems*. London, UK: International Maritime Organization.

[148] US-EPA (2003) *Ambient aquatic life water quality criteria for tributyltin (TBT) - Final*. Washington D.C., US.

[149] Anonymous (2000) *Waterbeleid voor de 21e eeuw - Geef water de ruimte en de aandacht die het verdient (Water policy in the 21st century - Give water the room and attention it deserves)*. Commissie

Waterbeheer 21e eeuw (Committee on watermanagement 21st century).

[150] Kolkman, M.J., van der Veen, A., Geurts, P.A.T.M. (2007) Controversies in water management: Frames and mental models. *Environmental Impact Assessment Review* doi: 10.1016/j.ciar.2007.05.005.

[151] Dewulf, A., Craps, M., Bouwen, R., Taillieu, T., Pahl-Wostl, C. (2005) Integrated management of natural resources: dealing with ambiguous issues, multiple actors and diverging frames. *Water Science and Technology* 52(6):115-24.

[152] Swaffield, S. (1998) Frames of reference: A metaphor for analysing and interpreting attitudes of environmental policy makers and policy influencers. *Environmental Management* 22(4):495-504.

[153] Tábábara, J.D. (2005) *Sustainability learning for River Basin Management and Planning in Europe.*

[154] Kolkman, M.J., Kok, M., van der Veen, A. (2005) Mental model mapping as a new tool to analyse the use of information in decision-making in integrated water management. *Physics and Chemistry of the Earth* 30:317-32.

[155] Newson, M. (2000) Science and sustainability: addressing the world water 'crisis'. *Progress in Environmental Science* 2(3):204-28.

[156] Rivett, P. (1994) *The craft of decision modelling*. Chichester, West Sussex, UK: John Wiley & Sons Ltd.

[157] Regier, H.A., Bronson, E.A. (1992) New perspectives on sustainable development and barriers to relevant information. *Environmental Monitoring and Assessment* 20:111-20.

[158] Haas, P.M. (1992) Introduction: Epistemic Communities and International Policy Coordination. *International Organization* 46:1-37.

[159] WorldBank (2002) *Sustainable development in a dynamic world; transforming institutions, growth, and quality of life*. Washington, U.S.A.: The International Bank for Reconstruction and Development/ The World Bank.

[160] Downs, A. (1972) Up and down with ecology - the issue-attention cycle. *The Public Interest* 28:38-50.

[161] Jaspers, F.G.W. (2003) Institutional arrangements for integrated river basin management. *Water Policy* 5:77-90.

[162] Gyawali, D., Allan, J.A., Antunes, P., Dudeen, A., Laureano, P., Luiselli Fernand,z, C., et al. (2006) *EU-INCO water research from FP4 to FP6 (1994-2006) - A critical review*. Brussels, Belgium: Directorate-General for Research and International Scientific Cooperation.

[163] SEPA (2001) *Bridging the Gap—Sustainability research and sectoral integration*. Stockholm, Sweden: SEPA.

[164] Brachet, C., Valensuela, D. (2012) *The handbook for integrated water resources management in transboundary basins of rivers, lakes and aquifers*. Paris, France: International Network of Basin Organizations - INBO.

[165] Medema, W., McIntosh, B.S., Jeffrey, P. (2007) Observations on fulfilling the promise of the IWRM and Adaptive Management concepts. *Ecology and Society*.

[166] Pahl-Wostl, C. (2007) Transition towards adaptive management of water facing climate and global change. *Water Resources Management* 21(1):49-62.

[167] Mysiak, J., Henriksen, H.J., Sullivan, C., Bromley, J., Pahl-Wostl, C. (2010) *The adaptive Water resource Management Handbook*. London, UK.: Earthscan.

[168] Hisschemöller, M., Groenewegen, P., Hoppe, R., Midden, C.J.H. (1998) *Kennisbenutting en politieke keuze: een dilemma voor het milieubeleid? (Use of knowledge and political choice: a dilemma for environmental policy?)*. The Hague, The Netherlands: Rathenau Instituut.

[169] Hisschemöller, M., Hoppe, R. (1995) Coping with intractable controversies: the case for problem structuring in policy design and analysis. 8(4):40-60.

[170] Hisschemöller, M., Timmerman, J.G., Langaas, S. (2004) Integrated assessment in transboundary water management: a tentative framework. *Environmental information in European transboundary water management*. London, UK: IWA Publishing. p. 168-83.

[171] RMNO (2005) *Interdisciplinarity and policy relevancy (Interdisciplinariteit en beleidsrelevantie)*. Den Haag, Netherlands: Raad voor Ruimtelijk, Milieu- en Natuuronderzoek.

[172] Kaiser, M. Multistakeholder application of the precautionary principle: the importance of transparent values. *Conference Multistakeholder application of the precautionary principle: the importance of transparent values*, Copenhagen, Denmark vol. International symposium on Uncertainty and Precaution in Environmental Management.

[173] Sanderson, H., Solomon, K.R. (2003) Precautionary limits to environmental science and risk management - three types of errors. *The Journal of Transdisciplinary Environmental Studies* 2(1):1-5.

[174] Koppejan, J.M.F., Klijn, E.H. (2004) *Managing uncertainties in networks*. London, UK: Routledge.

[175] Schindlmayr, T. (2001) The media, public opinion and population assistance: establishing the link. *International Family Planning Perspectives* 27(1):42-6.

[176] Agrawala, S. (1999) Early science-policy interactions in climate change: lessons from the Advisory Group on Greenhouse Gases. *Global Environmental Change* 9:157-69.

[177] Scheffer, M., Westley, F., Brock, W. (2003) Slow response of societies to new problems: causes and costs. *Ecosystems* 6:493-502.

[178] Cofino, W.P. (1995) Quality management of monitoring programmes In: Adriaanse, M., van de Kraats, J., Stoks, P.G., Ward, R.C. editors. *Conference Quality management of monitoring programmes*, Lelystad, The Netherlands. RIZA, p. 178-87.

[179] UNECE-TFMA (2000) *Guidelines on monitoring and assessment of transboundary rivers*. Lelystad, The Netherlands: UNECE Task Force on Monitoring and Assessment. RIZA.

[180] Timmerman, J.G., Bernardini, F. (2008) *Adapting to climate change in transboundary water management*. Perspective document for World Water Forum 5, Istanbul, Turkey: Co-operative Programme on Water and Climate (CPWC), the International Water Association (IWA), IUCN and the World Water Council.

[181] Raadgever, G.T., Mostert, E., Kranz, N., Interwies, E., Timmerman, J.G. (2008) Adaptive management of transboundary river basins - analysis of transboundary regimes from a normative perspective. *Ecology and Society*.

[182] Timmerman, J.G., Koeppel, S., Bernardini, F., Buntsma, J.J. (2011) Adaptation to Climate Change: Challenges for Transboundary Water Management. In: Leal Filho, W., editor. *The Economic, Social and Political Elements of Climate Change, Climate Change Management, Part 4* p. 523-41.

[183] van de Kerkhof, M., Huitema, D. (2004) Public participation in river basin management: A methodological perspective In: Timmerman, J.G., Behrens, H.W.A., Bernardini, F., Daler, D., Ross, P., Ruiten, C.J.M., et al. editors. *Conference Public participation in river basin management: A methodological perspective*, Lelystad, The Netherlands. RIZA/IWAC/RIKZ, p. 141-8.

[184] Turner, R.K., Timmerman, J.G., Langaas, S. (2004) Environmental information for sustainability science and management. *Environmental information in European transboundary water management*. London, UK: IWA Publishing. p. 153-67.

[185] Bouwen, R., Taillieu, T. (2004) Multi-party collaboration as social learning for interdependence: Developing relational knowing for sustainable natural resource management. *Journal of Community & Applied Social Psychology* 14:137-53.

[186] Mostert, E., Pahl-Wostl, C., Rees, Y.J., Searle, B., D. Tàbara, J.D., Tippett, J. (2007) Social learning in European river-basin management: Barriers and fostering mechanisms from 10 river basins. *Ecology and Society* 12(1):art. 19.

[187] Lowe, I. (2002) *The need for environment literacy*.

[188] Capurro, R., Fleissner, P., Hofkirchner, W. (1999) Is a unified theory of information feasible In: Hofkirchner, W. editor. *Conference Is a unified theory of information feasible*, Amsterdam, Netherlands vol. The quest for a unified theory of information. Gordon and Breach Publ., p. 9-30.

[189] Grossmann, M. (2005) *Kooperation an Afrikas internationalen Gewässern: die Bedeutung des Informationsaustauschs (Cooperation in Africa's transboundary waters: The role of information exchange)*. Bonn, germany: Deutsches Institut f r Entwicklungspolitik.

[190] Gudmundsson, H. (2003) The policy use of environmental indicators - Learning from evaluation research. *The Journal of Transdisciplinary Environmental Studies* 2(2):1-12.

[191] Kirschenbaum, M.G. (1998) *A white paper on information*.

[192] Scott, A. (2000) *The dissemination of the results of environmental research*.

[193] Nonaka, I., Takeuchi, H. (1995) *The knowledge-creating company. How Japanese companies create the dynamics of innovation*. New York: Oxford University Press.

[194] Roll, G., Timmerman, J.G., Langaas, S. (2004) Generation of usable knowledge in implementation of the European water policy. *Environmental information in European transboundary water management*. London, UK: IWA Publishing. p. 30-43.

[195] Romaldi, V. Collaborative technologies for knowledge management: making the tacit explicit? *Conference Collaborative technologies for knowledge management: making the tacit explicit?*, Cork, Ireland. p. 1357-65.

[196] Vlachos, E. Transboundary water conflicts and alternative dispute resolution. *Conference Transboundary water conflicts and alternative dispute resolution*. IWRA.

[197] Visscher, J.T., Pels, J., Markowski, V., de Graaf, S. (2006) *Knowledge and information management in the water and sanitation sector: a hard nut to crack*. Delft, The Netherlands: IRC International Water and Sanitation Centre.

[198] Hansen, A. (1991) The media and the social construction of the environment. *Media, Culture and Society* 13:443-58

[199] Shannon, C.E. (1948) A mathematical theory of communication. *The Bell System Technical Journal* 27:379-423.

[200] Nauen, C.E. (2005) *Increasing impact of the EU's international S&T sooperation for the transition towards sustainable development*. Luxembourg, Office for Official Publications of the European Communities.

[201] Braman, S. (1989) Defining information - an approach for policy-makers. *Telecommunications Policy* 13(3):233-42.

[202] Rowley, J. (1998) What is information? *Information Services & Use* 18:243-54.

[203] Capurro, R. (2001) Informationsbegriffe und ihre bedeutungsnetze. *Ethik und Sozialwissenschaften, Streitforum für Erwägungskultur* 12(1):14-7.

[204] Capurro, R. (2003) The concept of information. *Annual Review of Information Science and Technology* 37:343-411.

[205] Roll, G., Timmerman, J.G. (2006) Communicating with stakeholders and the public. In: Gooch, G.D., Stalnacke, P., editors. *Integrated transboundary water management in theory and practice: Experiences from the new EU Eastern borders*. London, UK: IWA Publishing. p. 127-48.

[206] Doody, J.P., Pamplin, C.F., Gilbert, C., Bridge, L. (1998) *Information required for Integrated Coastal Zone Management*. European Union Demonstration programme on integrated management in coastal zones.

[207] Timmerman, J.G., Langaas, S. (2004) Incorporating user needs into environmental information systems. *Environmental information in European transboundary water management*. London, UK: IWA Publishing. p. 108-24.

[208] Timmerman, J.G., Gooch, G.D., Kipper, K., Meiner, A., Mol, S., Nieuwenhuis, D., et al. (2003) The use and valuing of environmental information in the decision making process: an experimental study In: Bernardini, F., Landsberg-Uczciwek, M., Haunia, S., Adriaanse, M., Enderlein, R.E. editors. *Conference The use and valuing of environmental information in the decision making process: an experimental study*. 2003, p. 177-86.

[209] Mostert, E. (2005) *How can international donors promote transboundary water management?* Bonne, Germany: Deutsches Institut f r Entwicklungspolitik.

[210] Enderlein, R.E. (1999) Paper presented at the UN/ECE Working Group on Water Management meeting held on 15 September 1999 in Bonn, Germany. *Conference Paper presented at the UN/ECE Working Group on Water Management meeting held on 15 September 1999 in Bonn, Germany.*

[211] da Silva, J.E., Correia, F.N., da Silva, M.C. (1998) Transboundary issues in water resources. *Selected issues in water resources management in Europe Volume 2*. Rotterdam, The Netherlands: Balkema. p. 105-41.

[212] Savenije, H.G., van der Zaag, P. (2000) Conceptual framework for the management of shared river basins; with special reference to the SADC and EU. *Water Policy* 2:9-45.

[213] Huisman, P., de Jong, J., Wieriks, J.P. (2000) Transboundary cooperation in shared river basins: experiences from the Rhine, Meuse and North Sea. *Water Policy* 2(1-2):83-97.

[214] de Boer, J., Hisschemöller, M. (1999) *Concretisering informatie-behoefte: de diagnosefase (Concretising information needs: the phase of diagnosis).*

[215] Falkenmark, M., Tropp, H. (2005) Ecosystem approach and governance: contrasting interpretations. *Stockholm Water Front* 4:4-5.

[216] Gooch, G.D., Timmerman, J.G., Langaas, S. (2004) The communication of scientific information in institutional contexts: The specific case of transboundary water management in Europe. *Environmental information in European transboundary water management.* London, UK: IWA Publishing. p. 13-29.

[217] Moberg, F., Galaz, V. (2005) *Resilience: Going from conventional to adaptive freshwater management for human and ecosystem compatibility.* SIWI, Stockholm, Sweden.

[218] Ross, S.S. (2001) Muddy perceptions/dirty water: Messages for clearing the visions of the public and the powerful. *Conference Muddy perceptions/dirty water: Messages for clearing the visions of the public and the powerful*, Stockholm, Sweden. SIWI, p. 351-2.

[219] Serageldin, I., Steer, A. (1994) Making development sustainable. *Making development sustainable: from concepts to actions.* Washington D.C., USA: The World Bank. p. 1-3.

[220] van der Werff, P., Vellinga, P., van Drunen, M. (1999) Cultural evolution. *The environment A multidisciplinary concern.* Amsterdam, The Netherlands: Institute for Environmental Studies. p. 1-17.

[221] Winder, N. (2001) *What is 'Inter-Disciplinarity'?*

[222] Hoppe, R. (2005) Rethinking the science-policy nexus: from knowledge utilization and science technology studies to types of boundary arrangements. *PoiŠsis and Praxis: International Journal of Technology Assessment and Ethics in Science* 3(3):199-215.

[223] Brooks, N. (2003) *Vulnerability, risk and adaptation: A conceptual framework.* Tyndall Centre Working Paper No.38.

[224] Scott, A. (2001) Bridging the Gap between research and policy: sustainability research and sectoral integration. Rapporteur's report. *Conference Bridging the Gap between research and policy: sustainability research and sectoral integration. Rapporteur's report.*

[225] Sutherland, W.J., Armstrong-Brown, S., Armsworth, P.R., Brereton, T., Brickland, J., Campbell, C.D., et al. (2006) The identification of 100 ecological questions of high policy relevance in the UK. *Journal of Applied Ecology* 43(4):617-27.

[226] Timmerman, J.G., Langaas, S. (2005) Water information - what is it good for? On the use of information in transboundary water management. *Regional Environmental Change* 5(4):177-87.

[227] Takahashi, K., de los Angeles, M., Kuylenstierna, J. (2002) Workshop 2 (synthesis): driving forces and incentives for change towards sustainable water development. *Water Science and Technology* 45(8):141-4.

[228] Woodhill, A.J., Timmerman, J.G., Langaas, S. (2004) Dialogue and transboundary water resources management: towards a framework for facilitating social learning. *Environmental information in European transboundary water management*. London, UK: IWA Publishing. p. 44-59.

[229] McDaniels, T.L., Gregory, R. (2004) Learning as an objective within a structured risk management decision process. *Environmental Science and Technology* 38(7):1921-6.

[230] Mostert, E. (2004) Public participation and social learning for river basin management In: Timmerman, J.G., Behrens, H.W.A., Bernardini, F., Daler, D., Ross, P., Ruiten, C.J.M., et al. editors. *Conference Public participation and social learning for river basin management*, Lelystad, The Netherlands. RIZA / IWAC / RIKZ, p. 103-10.

[231] Bardwell, L.V. (1991) Problem-framing: a perspective on environmental problem-solving. *Environmental Management* 15(5):603-12.

[232] Gregory, R. (2000) Using stakeholder values to make smarter environmental decisions. *Environment* 42(5):34-44.

[233] Montazemi, A.R., Conrath, D.W. (1986) The use of cognitive mapping for information requirements analysis. *MIS Quarterly* 10(1):45-56.

[234] Keeney, R.L. (1992) *Value-focused thinking*. Cambridge, Massachusetts, USA: Harvard University Press.

[235] de Bono, E. (1979) *Lateral thinking*.

[236] Gooch, G.D. (2001) *Personal communication*.

[237] Keeney, R.L., Raiffa, H. (1993) *Decisions with multiple objectives: Preferences and value tradeoffs*. New York Cambridge University Press.

[238] Hogervorst, D. The Rhine action programme In: Walley, W.J., Judd, S. editors. *Conference The Rhine action programme*, Birmingham, UK. p. 35-42.

[239] Timmerman, J.G., Gardner, M.J., Ravenscroft, J.E. (1996) *Quality assurance*. Lelystad, The Netherlands: RIZA.

[240] Anonymous (2003) *Nationaal Bestuursakkoord Water (National administrative agreement on Water)*

[241] Lee, K.N. (1999) Appraising adaptive management. *Conservation Ecology* 3(2):3.

[242] Parr, T.W., Sier, A.R.J., Battarbee, R.W., Mackay, A., Burgess, J. (2003) Detecting environmental change: science and society. perspectives on long-term research and monitoring in the 21st century. *The Science of the Total Environment* 310:1-8

[243] MacDonald, L.H., Smart, A. (1993) Beyond the guidelines: practical lessons for monitoring. *Environmental Monitoring and Assessment* 26:203-18.

[244] Giordano, R., G., P., Barca, E. (2011) Monitoring Information Systems to Support Adaptive Water Management. In: Ekundayo, E.O., editor. *Environmental Monitoring*: Intech. p. 427-44.

[245] Nixon, S.C., Grath, J., B›gestrand, J. (1998) *EUROWATERNET. The European Environment Agency's Monitoring and Information Network for Inland Water Resources. Technical Guidelines for Implementation*. Copenhagen, Denmark: EEA.

[246] Helmer, R. Groundwater quality monitoring strategies In: Landsberg-Uczciwek, M., Adriaanse, M., Enderlein, R.E. editors. *Conference Groundwater quality monitoring strategies*, Mrzezyno, Poland. p. 285-98.

[247] Harrison, J.E. Key water quality monitoring questions: Designing monitoring and assessment systems to meet multiple objectives. *Conference Key water quality monitoring questions: Designing monitoring and assessment systems to meet multiple objectives*, Washington DC, USA. U.S. Environmental Protection Agency, p. 175-88.

[248] Mäkelä, A., Meybeck, M. (1996) Designing a monitoring programme. In: Bartram, J., Ballance, R., editors. *Water quality monitoring - A practical guide to the design and implementation of freshwater quality studies and monitoring programmes*. London, UK: Chapman & Hall. p. 35-59.

[249] Ongley, E.D., Ordoñez, E.B. (1997) Redesign and modernization of the Mexican water quality monitoring network. *Water International* 22:187-94.

[250] Sherwani, J.K., Moreau, D.H. (1975) *Strategies for water quality monitoring*. Springfield, VA, USA: North Carolina Water resources Research Institute.

[251] de Jong, J., Timmerman, J.G. (1997) Opening and introduction In: Ottens, J.J., Claessen, F.A.M., Stoks, P.G., Timmerman, J.G., Ward, R.C. editors. *Conference Opening and introduction*, Lelystad, The Netherlands. RIZA, p. 1-4.

[252] Timmerman, J.G., Beinat, E., Termeer, C.J.A.M., Cofino, W.P. (2010) Specifying information needs for Dutch national policy

evaluation. *Journal of Environmental Monitoring* (DOI: 10.1039/C0EM00135J).

[253] Bosch, O.J.H., Allen, W.J., Gibson, R.S. Monitoring as an integral part of management and policy making. *Conference Monitoring as an integral part of management and policy making*, New Zealand. Lincoln University, p. 12-21.

[254] Buishand, T.A., Hooghart, J.C. (1986) *Design aspects of hydrological networks*. The Hague, The Netherlands: TNO Committee on Hydrological Research; no. 35.

[255] Mulder, B.S., Noon, B.R., Spies, T.A., Raphael, M.G., Palmer, C.J., Olsen, A.R., et al. (1999) *The strategy and design of the effectiveness monitoring program for the Northwest Forest Plan*. Portland, OR: U.S.: Department of Agriculture, Forest Service, Pacific Northwest Research Station.

[256] Passino, R., Seager, J. A manual of best practice for water monitoring: UK-Italy collaborative programme In: Ottens, J.J., Claessen, F.A.M., Stoks, P.G., Timmerman, J.G., Ward, R.C. editors. *Conference A manual of best practice for water monitoring: UK-Italy collaborative programme*, Lelystad, The Netherlands. RIZA, p. 21-34.

[257] Cofino, W.P. (1989) Methodology of chemical monitoring in the marine environment. *Helgoland Marine Research* 43(3):295-308.

[258] Broderick, B.E., Cofino, W.P., Cornelis, R., Heydorn, K., Horwitz, W., Hunt, D.T.E., et al. (1991) A journey through quality control. *Mikrochimica Acta* II:523-42.

[259] Wells, D.E., Cofino, W.P., Quevauviller, P., Griepink, B. (1993) Quality assurance of information in marine monitoring. *Marine Pollution Bulletin* 26(7):368-75.

[260] Ward, R.C. (1995) Water quality monitoring as an information system In: Adriaanse, M., van de Kraats, J., Stoks, P.G., Ward, R.C. editors. *Conference Water quality monitoring as an information system*, Lelystad, The Netherlands. RIZA, p. 84-92.

[261] McDonnell, R.A. (2008) Challenges for Integrated Water Resources Management: How do we provide the knowledge to support truly integrated thinking? *Water Resources Development* 24(1):131-43.

[262] Stroomberg, G.J., Freriks, I.L., Smodos, F., Cofino, W.P., Quevauviller, P. (1995) Quality assurance and quality control of surface water sampling. *Quality Assurance in Environmental Monitoring: Sampling and Sample Pretreatment*. Weinheim, Germany: Wiley-VCH.

[263] Timmerman, J.G., Hendriksma, J. (1997) Informatie op maat: een raamwerk voor waterbeheer (Tailor-made information: a framework for water management). *H2O* 30(17):582-30.

[264] Timmerman, J.G., Ottens, J.J., Ward, R.C. (2000) The information cycle as a framework for defining information goals for water-quality monitoring. *Environmental Management* 25(3):229-39.

[265] WCMC (1998) *Volume 6: Information management capacity*. London, UK: World Conservation Monitoring Centre. Commonwealth secretariat.

[266] CIW (2001) *Leidraad monitoring. Definitief rapport (Instructions for monitoring. Final report)* Netherlands: Commission on Integrated Water Management.

[267] Gilde, L.J., Prins, K.H., Helmond, C.A.M. (1999) *Monitoring zoete rijkswateren (Monitoring of the national inland waters)*

[268] MinV&W (2007) *Watermarkt > Meetprocessen > De informatiekring-loop*.

[269] Dannisoe, J.G., Larsen, H. The use of models in monitoring strategies In: Ottens, J.J., Claessen, F.A.M., Stoks, P.G., Timmerman, J.G., Ward, R.C. editors. *Conference The use of models in monitoring strategies*, Lelystad, The Netherlands. RIZA, p. 247-52.

[270] Pasche, E. Monitoring and assessment in rivers based on two-dimensional models In: Ottens, J.J., Claessen, F.A.M., Stoks, P.G., Timmerman, J.G., Ward, R.C. editors. *Conference Monitoring and assessment in rivers based on two-dimensional models*, Lelystad, The Netherlands. RIZA, p. 263-76.

[271] Schulze, F.H., Bouma, N.A. Use of artificial neural networks in integrated water management In: Timmerman, J.G., Cofino, W.P., Enderlein, R.E., J lich, W., Literathy, P.L., Martin, J.M., et al. editors. *Conference Use of artificial neural networks in integrated water management*, Lelystad, The Netherlands. RIZA/IWAC, p. 333-42.

[272] Sanders, T.G., Loftis, J.C. Factors to consider in optimization of a monitoring network In: Adriaanse, M., van de Kraats, J., Stoks, P.G., Ward, R.C. editors. *Conference Factors to consider in optimization of a monitoring network*, Lelystad, The Netherlands. RIZA, p. 146-52.

[273] Cofino, W.P., Barcel¢, D. (1993) Quality assurance in environmental analysis. *Environmental analysis: Techniques, applications and quality assurance*. Amsterdam, The Netherlands: Elsevier Science Publishers. p. 359-81.

[274] UNECE-TFMA (1996) *Guidelines on water-quality monitoring and assessment of transboundary rivers*. Lelystad, The Netherlands: UNECE Task Force on Monitoring and Assessment.

[275] Ward, R.C. (1986) Framework for designing water quality information systems In: Lerner, D. editor. *Conference Framework for designing water quality information systems*, Wallingford, UK

vol. IAHS publication 157. International Association of hydrological Sciences, p. 89-98.

[276] Latour, P.J.M., Stutterheim, E., Schäfer, A.J. From data to information: the Water Dialogue In: Ottens, J.J., Claessen, F.A.M., Stoks, P.G., Timmerman, J.G., Ward, R.C. editors. *Conference From data to information: the Water Dialogue*, Lelystad, The Netherlands. RIZA, p. 481-8.

[277] Klapwijk, S.P., Gardeniers, J.J.P., Peeters, E.T.H.M., Roos, C. Ecological assessemnt of water systems In: Adriaanse, M., van de Kraats, J., Stoks, P.G., Ward, R.C. editors. *Conference Ecological assessemnt of water systems*, Beekbergen, The Netherlands. RIZA, p. 105-17.

[278] Ten Brink, B.J.E., Hosper, S.H., Colijn, F. (1991) A quantitative method for description and assessment of ecosystems: The AMOEBE approach. *Marine Pollution Bulletin* 23:265-70.

[279] McBride, G.B., Smith, D.G. (1997) Results of a trend assessment of New Zealand's National River Water Quality Network In: Ottens, J.J., Claessen, F.A.M., Stoks, P.G., Timmerman, J.G., Ward, R.C. editors. *Conference Results of a trend assessment of New Zealand's National River Water Quality Network*, Lelystad, The Netherlands. RIZA, p. 135-42.

[280] UNECE (1996) *Protection of transboundary waters, guidance for policy- and decision-making*. Geneva, Switzerland: UN.

[281] Ward, R.C., Timmerman, J.G., Peters, C.A., Adriaanse, M. (2004) In search of a common water quality monitoring framework and terminology In: Timmerman, J.G., Behrens, H.W.A., Bernardini, F., Daler, D., Ross, P., Ruiten, C.J.M., et al. editors. *Conference In search of a common water quality monitoring framework and terminology*, Lelystad, The Netherlands. RIZA / IWAC / RIKZ, p. 195-206.

[282] Smith, D.G. (2004) Designing a complex multi-objective water quality monitoring network: The New York City water supply example. *Conference Designing a complex multi-objective water quality monitoring network: The New York City water supply example*. NWQMC.

[283] Mintzberg, H., Ahlstrand, B., Lampel, J. (1998) *Strategy safari. A guided tour through the wilds of strategic management*. London, UK: Prentice Hall.

[284] Hopstaken, B., Kranendonk, A. (1991) *Informatieplanning: puzzelen met beleid en plan. (Information-planning: puzzling over policy and plan)* Deventer, The Netherlands: Kluwer.

[285] Meppem, T., Gill, R. (1998) Planning for sustainability as a learning concept. *Ecological Economics* 26:121-37.

[286] Checkland, P.B. (1981) *Systems thinking, systems practice.* Chichester, England: John Wiley & Sons.

[287] Chiew, V., Khan, S., Man, J. (1999) *Further methodologies for requirements engineering: SSM Web Report.*

[288] Finegan, A. (1994) Soft Systems Methodology: An Alternative Approach to Knowledge Elicitation in Complex and Poorly Defined Systems. *Complexity International* 1(April 1994).

[289] Tsouvalis, C., Checkland, P.B. (1996) Refelecting on SSM: the dividing line between 'real world' and 'systems thinking world'. *Systems Research* 13(1):35-45.

[290] Shehata, M., Bowen, S. (2003) *Soft Systems Methodology.*

[291] Bunch, M.J. (2003) Soft Systems Methodology and the Ecosystem Approach: A system study of the Cooum River and environs in Chennai, India. *Environmental Management* 32(2):182-97.

[292] Couprie, D., Goodbrand, A., Li, B., Zhu, D. (1997) *Soft Systems Methodology.*

[293] Argelo, S., Boterman, J. (1991) *Praktijkboek informatieplanning; opbrengsten en werkwijzen (Practical book on information planning; profits and procedures).* Deventer, The Netherlands: Stenfert Kroese.

[294] Belton, V., Stewart, T.J. (2002) *Multiple criteria decision analysis. An integrated approach.* Boston/Dordrecht/London: Kluwer Academic Publishers.

[295] Mingers, J., Rosenhead, J. (2004) Problem structuring methods in action. *European Journal of Operational Research* 152(3):530 - 54.

[296] Karstens, S. Scale choices in policy support for water management In: Timmerman, J.G., Behrens, W., Bernardini, F., Turner, R.K. editors. *Conference Scale choices in policy support for water management.* RIZA/IWAC/RIKZ, p. 215-24.

[297] Harremoës, P., Turner, R.K. (2001) Methods for integrated assessment. *Regional Environmental Change* 2(2):57-65.

[298] Yourdon, E. (1989) *Modern structured analysis*: Prentice-Hall, Englewood Cliffs, NJ, USA.

[299] Jain, R.K., Urban, L.V., Stacey, G.S., Balbach, H.E. (1993) *Environmental assessment*: McGraw-Hill, inc., USA.

[300] Ridder, D., Mostert, E., Wolters, H.A. (2005) *Learning together to manage together - Improving participation in water management.* Osnabr ck, Germany: Druckhaus Bergmann.

[301] Timmerman, J.G., de Boer, J., Hisschemöller, M., Mulder, W.H. (2001) Specifying information needs: improving the working methodology. *Regional Environmental Change* 2:77-84.

[302] Schwenk, C. (1984) Devil's advocate in managerial decision making. *Journal of Management Studies* 21:153-68.

[303] Holmberg, J., RobŠrt, K.H. (2000) Backcasting from non-overlapping sustainability principles - a framework for strategic planning. *International Journal of Sustainable Development and World Ecology* 7:291-308.

[304] Turner, W.S., Langerhorst, R.P., Hice, G.F., Eilers, H.B., Uijttenbroek, A.A. (1988) *System development methodology.* Nijmegen, The Netherlands: Thieme bv.

[305] Adriaanse, M. (1995) Information requirements as design criteria for surface water monitoring In: Adriaanse, M., van de Kraats, J., Stoks, P.G., Ward, R.C. editors. *Conference Information requirements as design criteria for surface water monitoring*, Beekbergen, The Netherlands. RIZA, p. 126-33.

[306] UNECE (2011) *Second assessment of transboundary rivers, lakes and groundwaters.* Geneva, Switzerland: United Nations.

[307] Landsberg-Uczciwek, M. (2002) *Bug: Identification and review of water management issues. Report No. 2.* Szczecin, Poland: RIZA.

[308] UNECE (1992) *Convention on the Protection and Use of Transboundary Water Courses and International Lakes.* Helsinki, Finland.

[309] Burchi, S., Mechlem, K. (2005) *Groundwater in international law. Compilation of treaties and other legal instuments*: FAO/UNESCO.

[310] Timmerman, J.G., Beinat, E., Termeer, C.J.A.M., Cofino, W.P. (2010) A methodology to bridge the water information gap. *Water Science and Technology* 62(10):2419-26.

[311] Bana e Costa, C., Beinat, E. (2005) Model-structuring and impact assessment: qualitative value analysis of policy attractiveness.

[312] Keeney, R.L. (1994) Creativity in decision making with value-focused thinking. *Sloan Management Review* Summer 1994:33-41.

[313] Bernstein, B.B., Thompson, B.E., Smith, R.W. (1993) A combined science and management framework for developing regional monitoring objectives. *Coastal Management* 21:185-95.

[314] Timmerman, J.G., Mulder, W.H. (1999) Information needs as the basis for monitoring. *European Water Management* 2(2):41-5.

[315] WCMC (1998) *Volume 2: Information needs analysis.* London, UK: World Conservation Monitoring Centre. Commonwealth secretariat.

[316] GWP-INBO (2009) *A Handbook for Integrated Water Resources Management in Basins*: Global Water Partnership (GWP) and the International Network of Basin Organizations (INBO).

[317] Lobbrecht, A.H. (1997) *Dynamic water-system control design and operation of regional water-resources systems*: TU Delft, The Netherlands.

[318] Claassen, T.H.L. (2001) *Naar een monitoringprogramma voor het thema Gradiënten. Informatiebehoefte als basis voor informatiestrategie voor de randen van de Waddenzee (Towards a monitoring network for Gradients. Information needs as basis for an informationstrategie for the edges of the Wadden Sea)* Leeuwarden, The Netherlands: Rijkswaterstaat, Directie Noord-Nederland.

[319] Adriaanse, A. (1993) *Environmental policy performance indicators*. The Hague, The Netherlands: Sdu.

[320] Gallopín, G.C. (1996) Environmental and sustainability indicators and the concept of situational indicators. A systems approach. *Environmental Modelling and Assessment* 1:101-17.

[321] OECD (1993) *OECD core set of indicators for environmental performance reviews. A synthesis by the group on the state of the environment*. Paris, France: OECD.

[322] Ottens, J.J., Timmerman, J.G., van der Grift, B., Sprenger, T. (1998) Prospects for the use of indicators in water management In: Landsberg-Uczciwek, M., Adriaanse, M., Enderlein, R.E. editors. *Conference Prospects for the use of indicators in water management*, Mrzezyno,Poland. p. 311-23.

[323] Bakkes, J.A., van der Born, G.J., Helder, J.C., Swart, R.J. (1994) *An overview of environmental indicators: state of the art and perspectives*. Bilthoven, The Netherlands.

[324] Sullivan, C. (2002) Calculating a Water Poverty Index. *World Development* 30(7):1195-210.

[325] Schiller, A., Hunsaker, C.T., Kane, M.A., Wolfe, A.K., Dale, V.H., Suter, G.W., et al. (2001) Communicating ecological indicators to decision makers and the public. *Conservation Ecology* 5(1):19.

[326] Boulton, A.J. (1999) An overview of river health assessment: philo-sophies, practice, problems, and prognosis. *Freshwater Biology* 41:469-79.

[327] Hammond, A., Adriaanse, A., Rodenburg, E., Bryant, D., Woodward, R. (1995) *Environmental indicators: A systematic approach to measuring and reporting on environmental policy performance in the context of sustainable development*. New York, USA: World Resources Institute.

[328] Smeets, E., Weterings, R. (1999) *Environmental indicators: Typology and overview*. Copenhagen, Denmark: EEA.

[329] van Harten, H.A.J., van Dijk, G.M., de Kruijf, H.A.M. (1995) *Waterkwaliteitsindicatoren: overzicht, methode-ontwikkeling en toepassing (Water quality indices: overview, methodologies and application)* Bilthoven, The Netherlands: RIVM.

[330] Friend, A., Rapport, D. (1979) *Towards a comprehensive framework for environment statistics: A stress-response approach*. Ottawa, Canada: Statistics Canada.

[331] EEA (1998) *Europe's environment: the second assessment*. Oxford, UK: Elsevier Science Ltd.

[332] Wieringa, K. (1996) Towards integrated environmental assessment supporting the community's environmental action programme process. *Conference Towards integrated environmental assessment supporting the community's environmental action programme process*.

[333] Anonymous (2001) *Morava river pilot project. Information needs report Slovak Republic*. Bratislava, Slovak Republic: Ministry of Environment/Slovak Hydrometeorological Institute.

[334] Durcovicova, D., Vydareny, M. (2001) *Latorica/Uh Rivers pilot project: Information needs report; Slovak Republic*. Bratislava, Slovak Republic: Slovak Hydrometeorological Institute.

[335] Kelnárová, Z., Vancová, A., Adamková, J., Sevciková, V. (2001) *Ipel'/Ipoly river pilot project. Information needs report Slovak Republic*. Bratislava, Slovak Republic: Ministry of Environment/ Slovak Hydrometeorological Institute.

[336] László, F. (2000) *Water quality monitoring of transboundary rivers: Ipel'/Ipoly River pilot project; Draft information needs report Hungary*. Budapest, Hungary: VITUKI.

[337] László, F. (2000) *Water quality monitoring of transboundary rivers: Mures/Maros River pilot project; Draft information needs report Hungary*. Budapest, Hungary: VITUKI.

[338] Landsberg-Uczciwek, M., Korol, R., Zan, T. (2001) *Information needs report; Pilot project of monitoring and assessment of water quality in the basin of the Bug River*.

[339] Ognean, R.C. (2000) *Water quality monitoring of transboundary rivers: Mures/Maros River pilot project; Draft information needs report: Information needs of the RO part of the river basin* Bucuresti, Romania: Romanian Waters Authority 'Apele Romane'.

[340] Inckel, M.S. (2002) *Evaluatie project Monitoring Maas; de laatste ronde (Evaluation project Monitoring River Meuse; the final round)*. Maastricht, The Netherlands: Rijkswaterstaat Regional Department of Limburg.

[341] Rijkswaterstaat (2001) *Monitoringstrategie Maas. Fase 3: Inwin scenario's (Monitoring strategy river Meuse. Phase 3: Data collection scenario's)*. Maastricht, The Netherlands: Directoraat-Generaal Rijkswaterstaat, Directie Limburg / IWACO adviesbureau voor water en milieu.

[342] Witmer, M.C.H. Information needs for policy evaluation In: Adriaanse, M., van de Kraats, J., Stoks, P.G., Ward, R.C. editors. *Conference Information needs for policy evaluation*, Lelystad, The Netherlands. RIZA, p. 55-61.

[343] CIW (1996) *Voortgangsrapportage integraal waterbeheer en Noordzee aangelegenheden (Progress report on integrated water management and North Sea matters)* Netherlands: Commission on Integrated Water Management/CUWVO.

[344] Kristensen, P., Krogsgaard Jensen, J. (1997) Integrated approach for chemical, biological and ecotoxicological monitoring - a tool for environmental management In: Ottens, J.J., Claessen, F.A.M., Stoks, P.G., Timmerman, J.G., Ward, R.C. editors. *Conference Integrated approach for chemical, biological and ecotoxicological monitoring - a tool for environmental management*, Lelystad, The Netherlands. RIZA, p. 111-20.

[345] Stoks, P.G. (1995) Water quality control in the production of drinking water from river water In: Adriaanse, M., van de Kraats, J., Stoks, P.G., Ward, R.C. editors. *Conference Water quality control in the production of drinking water from river water*, Beekbergen, The Netherlands. RIZA, p. 118-25.

[346] Kleijberg, R., Verheij, H., Bertens, P. (2000) *Integraal Monitoringsplan Zandmaas/Maasroute (Integrated monitoring plan Zandmaas/Maasroute).* ‘s-Hertogenbosch, The Netherlands: ARCADIS Heidemij Advies BV.

[347] Rijkswaterstaat (2001) *Monitoringstrategie Maas. Fase 1: Input analyse (Monitoring strategy river Meuse. Phase 1: Input analysis).* Maastricht, The Netherlands: Directoraat-Generaal Rijkswaterstaat, Directie Limburg/IWACO adviesbureau voor water en milieu.

List of the Figures

List of the Tables

List of the Boxes

Index

T

V

W

Color Plate Section

Chapter 1

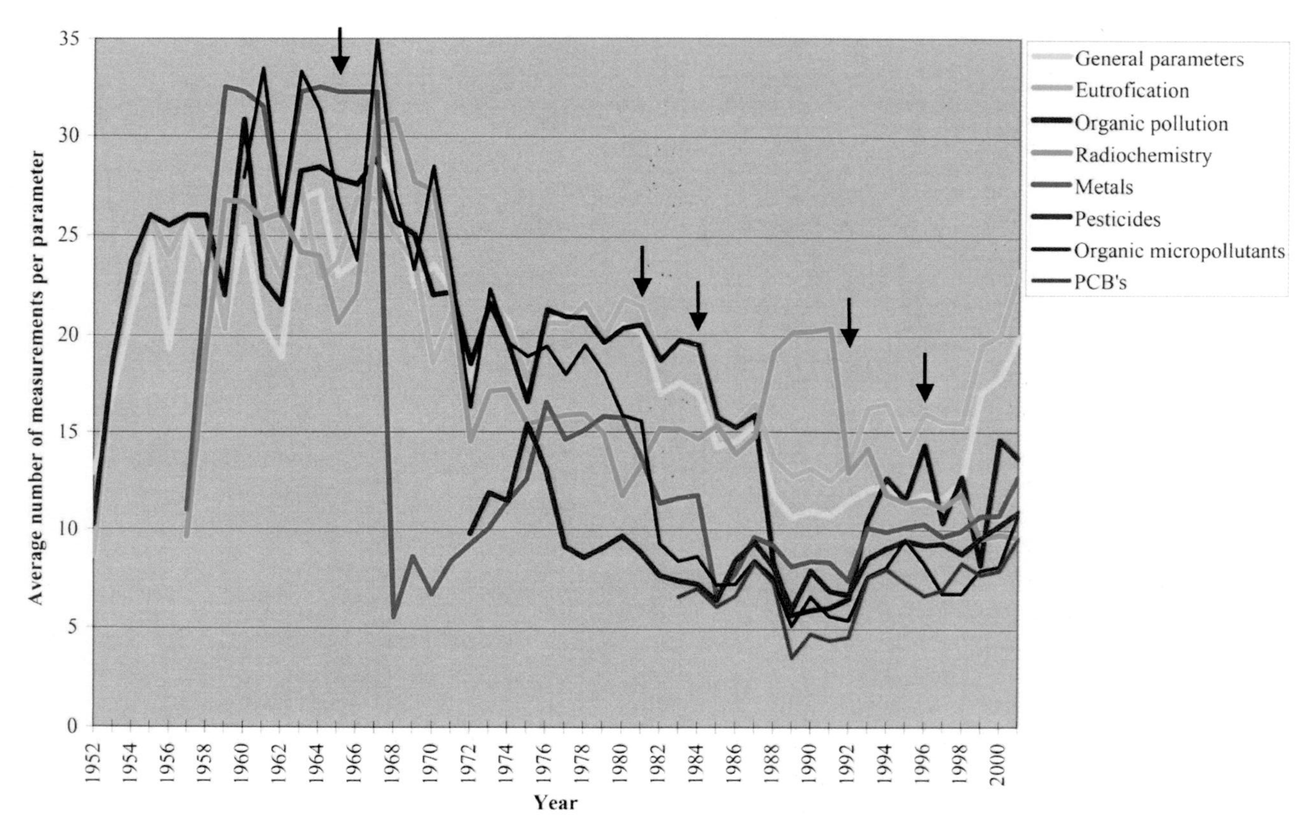

Figure 1.5 Average number of measurements per parameter per year for each parameter group (arrows indicate the years that evaluation studies were performed)

Chapter 2

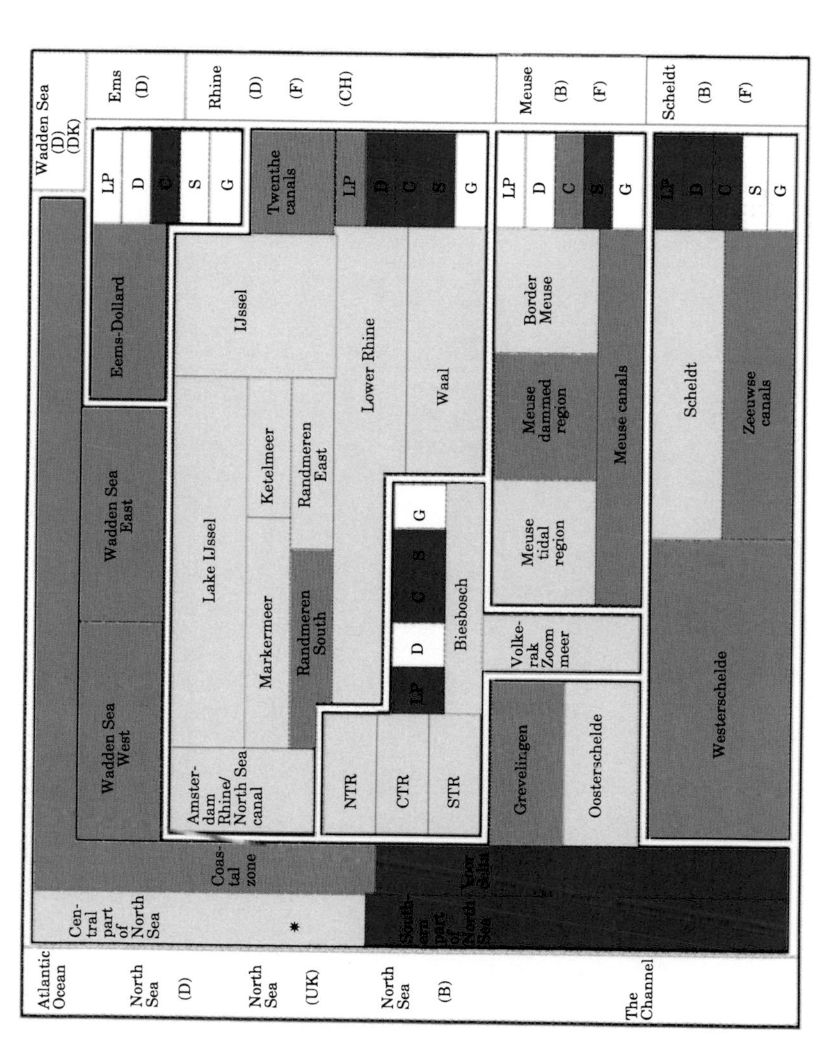

Figure 2.5 In the water dialogue software system, a schematic representation of the map of The Netherlands is presented called the Water Mondriaan. This graphical presentation shows the major water bodies. The color of each box represents the state of that water body[276].